# Space Odyssey

## Spatiality and Social Relations in the 21st Century

*Edited by*

JØRGEN OLE BÆRENHOLDT
KIRSTEN SIMONSEN
*Roskilde University, Denmark*

Routledge
Taylor & Francis Group

LONDON AND NEW YORK

First published 2004 by Ashgate Publishing

2 Park Square, Milton Park, Abingdon, Oxon OX14 4RN
711 Third Avenue, New York, NY 10017, USA

*Routledge is an imprint of the Taylor & Francis Group, an informa business*

First issued in paperback 2017

**British Library Cataloguing in Publication Data**
Space odysseys : spatiality and social relations in the
    21st century
    1.Human geography - Congresses 2.Geographical perception -
    Congresses 3.Spatial behavior - Congresses
    I. Bærenholdt, Jørgen Ole II. Simonsen, Kirsten
    304.2'3

**Library of Congress Cataloging-in-Publication Data**
Space odysseys : spatiality and social relations in the 21st century / by Jørgen Ole Bærenholdt and Kirsten Simonsen.
        p. cm.
    Includes bibliographical references and index.
    ISBN 0-7546-4349-2
    1. Human geography. 2. Human territoriality. 3. Spatial behavior. 4. Personal space. 5. Public spaces. I. Bærenholdt, Jørgen Ole. II. Simonsen, Kirsten, 1946-

    GF50.S64 2004
    304.2--dc22
                                                                2004013252
ISBN 978-0-7546-4349-4 (hbk)
ISBN 978-1-138-27652-9 (pbk)

Typeset in Times Roman by N$^2$productions

# Contents

# List of Figures

# List of Contributors

**Jörg Bechmann.** PhD in sociology at University of Copenhagen. Has worked as a transport researcher with the Institute for Urban and Regional Development in Dortmund, Germany, and the University of Copenhagen, Denmark. Books and articles displaying a social science perspective on mobility, transport and traffic. Involved in developing and promoting NGO-positions for a safer and more sustainable European transport policy.

**Keld Buciek.** PhD in geography at Roskilde University, Denmark, following studies at University of Copenhagen. Associate Professor in geography at Roskilde University. Fieldwork in Ghana and academic interest in travelling and mobility.

**Jørgen Ole Bærenholdt.** PhD in geography at Roskilde University, Denmark. Still there as associate professor and head of geography. Also affiliated with planning at the University of Tromsø, Norway. Fieldwork in Greenland, Iceland, Faroes, and Murmansk Region of Russia. Author with Michael Haldrup, Jonas Larsen and John Urry of *Performing Tourist Places* (2004).

**Derek Gregory.** PhD in geography at Cambridge University, UK and started his academic career in Cambridge. Professor in geography at University of British Columbia, Vancouver. Doctor of Honour, Roskilde University. Author of a number of seminal books including *Ideology, Science and Geography* (1978), and *Geographical Imaginations* (1994) and *The Colonial Present* (2004).

**Costis Hadjimichalis.** Department of Geography, Harokopio University, Athens. First degree in architecture at AUTH and later PhD in geography and planning at UCLA. Academic career at the Planning Department, AUTH until 1997 and visiting professor in European and USA universities. Research on economic geography, uneven regional development and geopolitics in Greece, the Balkans and Southern Europe.

**Ole B. Jensen.** PhD from Aalborg University, Denmark. Works as Associate Professor in 'Sociology and the Spatial Transformation Processes of Society' at Department of Development and Planning. Currently affiliated with University of Sheffield, UK and has previously been affiliated with University of Newcastle, UK.

**Tim Richardson.** Senior Lecturer in the Department of Town and Regional Planning, University of Sheffield and Visiting Professor in European Spatial Policy at Aalborg

University. PhD from Sheffield Hallam University and has worked as a researcher in Manchester, Leeds and Aberdeen universities. Researched policy making at the heart of the EU in Brussels; national parks in England and Scotland, local authorities in the North of England; and ideas of 'Europe', wherever they are encountered.

**Kirsten Simonsen.** PhD in geography from University of Copenhagen, Dr. Phil. Roskilde University. Professor in social geography, Roskilde University, Denmark. Has worked in Copenhagen University, Aarhus University, NORDPLAN in Stockholm, and among her publications is the reader *Voices from the North, New Trends in Nordic Human Geography* (edited with Jan Öhman, 2003).

**Anke Strüver.** Degrees in film studies, Ruhr University, Bochum, Germany, and geography, University of Hamburg, Germany. Worked at the Nijmegen Centre for Border Research, University of Nijmegen, the Netherlands as PhD student and lecturer, but now at the Department of Geography of the University of Münster, Germany. Academic focus on borders, identity and belonging as well as on popular culture and social theory in human geography.

**John Urry.** First degree in economics, PhD sociology, Cambridge University. Professor in sociology, Lancaster University. Visiting Professor and Doctor of Honour at Roskilde University. Author of long list of seminal books, including *Global Complexity* (2003), *The Tourist Gaze* (sec. ed. 2002), *Sociology Beyond Societies* (2000), *Consuming Places* (1995) and *Economies of Signs and Space* (with Scott Lash, 1994).

**Dina Vaiou.** PhD from the University of London. MA from UCLA. Associate Professor in the Department of Urban and Regional Planning of the National Technical University of Athens. Fieldwork in Athens, Northern Greece and Southern Europe.

**Benno Werlen.** Used to lecture in geography at the University of Zurich and the Swiss Federal Institute of Technology, Zurich. Now Professor in geography, Friedrich-Schiller-University of Jena, Germany. Author of *Society, Action and Space* (1993) and *Sozialgeographie Alltäglicher Regionalisierungen* [*Social Geography of Everyday Regionalizations*] vols I and II (1995 and 1997).

**Fiona Wilson.** PhD in geography from Cambridge University. 'Development' research with main focus on Latin America. Several periods of fieldwork in Peru. Used to be Senior Researcher at the Centre of Development Studies/Danish Institute of International Studies, Copenhagen, Denmark, but now back as Professor in International Development Studies, Roskilde University.

**Wolfgang Zierhofer.** Training in human geography at the University of Zurich, Switzerland. PhD in geography at the Federal Institute of Technology (ETH), Zurich. Worked as Associate Professor at the University of Nijmegen, Netherlands. Now research and teaching affiliations with the Program Man, Society and Environment at the University of Basel, Switzerland.

# Preface

This book presents a selection of papers from the international seminar *2001: A Space Odyssey* seminar held by the Geography Section of the Department of Geography and International Development Studies, Roskilde University, Denmark in November 2001. We are happy that the seminar received funding from the Social Science Research Council of Denmark. The seminar would never have taken place, without the practical, administrative, logistic and other support from Anette Skaarup and Annemette Porse. We would like to thank them for their efforts in making this seminar a rewarding experience. Also other colleagues in our department and many other participants made crucial efforts, helping to arrange the seminar and produce this publication. Thanks to Ashgate for assistance in formatting the text for publication, and thanks especially to Valerie Rose and Sarah Horsley for their efforts in producing the book.

The seminar took place only a few months after September 11, 2001. The crusades symbolised by Osama Bin Laden and George Bush formulated the agenda, although few at that time knew how much these crusades would change the world for the worse in the following years. Anxiety about the state of the world was always present, but this was counter-balanced by a desire to strengthen social relations and interdependencies, as a safeguard against ever present risks. Apart from the last chapter, this is not specifically a book about September 11, but the subsequent wars in Afghanistan and Iraq may have contributed in making the odyssey of this collection longer and more complex than first expected.

Indeed we tried to challenge the authors in the various rounds of editorial comments and re-writing. The flow of texts followed very different temporalities and we would like to thank the contributors for their patience and confidence in our efforts. We learned a lot and only hope that others will also feel inspired by reading this book.

Jørgen Ole Bærenholdt
Kirsten Simonsen

Roskilde, April 2004

# Introduction

Kirsten Simonsen and Jørgen Ole Bærenholdt

## From where we take off

Spatial questions are today on the social and political agenda and the nature of spatial imaginations has become central to a range of the major contemporary social debates. Narratives on spatial inequality, from the North-South divide in global economic and political visions to marginalisation and 'ghettoisation' in Western cities, appear regularly in our daily newspapers. Immigration from the third world countries underlines mobility in the contemporary world and challenges Western/European national identities. And 'globalisation' has become a mantra in economic and political discourse. At the same time, however, national, regional and local political projects show that spatial collective identities are not at all a vanishing phenomenon. Political fights in Seattle, Prague, Nice and Gothenburg have showed that fairly different spatial imaginations can be involved in the contested area of globalisation. These few examples should be enough to illustrate that issues of space/spatiality – an old geographical issue – are as crucial in our current societies as ever before. Saying this should not be seen as some kind of commitment to disciplinary boundaries or traditions. Theories of spatiality have been developed within geography and from there contributed to studies initiated from other perspectives. Also, geography has taken perspectives from other parts of academia. As it will appear from the contributions to this book, geography is seen as a discourse rather than a discipline and, subsequently, the whole issue of spatiality is considered a cross-disciplinary one.

Within this background we will initiate an (admittedly very partial) odyssey through the intellectual landscape of what we have called 'spatiality and social relations in the twenty-first century'. The journey takes off in the middle of the 1980s. At that time, a wave of theoretical discussions on socio-spatial relations culminated, among other places in collections in England (Gregory and Urry 1985, *Social Relations and Spatial Structures*), Denmark (Tonboe 1985, *Farvel til byen?*) and France (Auriac and Brunet 1986, *Espaces, jeux et enjeux*). The most important achievement gained from these debates is the, now fairly generally accepted, understanding that space (and time) are social constructs. This relational view of space prioritises analyses of how space is constituted and given meaning through human endeavour – how it is a product of cultural, social, political and economic relations. Space/spatiality is a *dimension* of social relations and imaginations; it is not an objective structure but rather a social experience. That does not mean that it is a mere social convention devoid of any material basis, rather it is a conception constructed by way of people's social practices in their involvement with the world.

One consequence of the understanding of space as a social construct is an interest in the way in which different social configurations produce qualitative different conceptions of space. Probably the most well-known account of that comes from Henri Lefebvre who in his *Production of Space* (1991/1974) frames a modern, capitalist spatiality by way of a history of space expounding a shift from absolute space, over historical space to abstract space. This narrative of denaturalisation and decorporealisation (cf. Gregory 1994) of space has been highly influential, and it has probably, directly or indirectly, left its mark on most contemporary discussions of space. It is within this background that the first aim of this book should be seen. We find it worthwhile, now a little more than 15 years later, to consider the course of the journey and resume the debate on socio-spatial relationships, taking on board new theoretical trends and new social circumstances. Much change has come about in this period, and it is worth reconsidering our conceptions of space and possible reasons for their reimagination.

Thus, the book starts from the presupposition that space/spatiality is a social dimension and a social construct. This is now more or less conventional wisdom, and so is the dual proposition that space is constituted by, as well as constitutive of, social relations and social practices. In practice, however, and often against theoretical pronouncements, we often experience a slippage into much more prosaic approaches in which social relations and social practices occur within some pre-constituted and static framework of space. Therefore, a second aim of the book is to present examples of these conceptions of space *in work*. This is both a theoretical and an empirical endeavour, it is about the ways in which a social understanding of spatiality can form the ground of additional conceptual development and grasp the complexity of the social relations we experience in the present world.

Using the indirect reference to Arthur T. Clarke's novel and Stanley Kubrick's science fiction movie *2001: A Space Odyssey* in the title of the book already suggests some of the changes that inform the conceptual reconstructions. The novel and the movie were a distopian imagination of the twenty-first century in which artificial intelligence takes control and ends human life. Together with its representation of HAL (the computer) as human-like reacting being, it is known for its dehumanised description of human characters and social contacts, reaching its climax in the principal character's escape from women and family ties represented by mother Earth. It can be argued, then, that the movie offers some clues to a (new) spatiality of the twenty-first century. Mobility and the voyage as a symbol of cultural identity is one of the perspectives to be read off, and so is standardisation of space through technology – to machine intelligence, man-machine hybridisation and dehumanisation of action.

By operating with *Space Odysseys* in plural, however, we try to suggest a less deterministic journey into the twenty-first century. That is, we want to open an exploration of the conception of space/spatiality consisting of multiple trajectories, some of which might co-exist, affect each other, or even come into conflict. The difference of these trajectories might reflect differences in experience, in social context and/or in gender, class and ethnicity. In particular, we want to stress that the production of knowledge itself is a spatial practice in terms of the situations, positions, networks, flows, connections, places and performances involved. Against universalistic claims on the character of spatiality and social relations in the twenty-first century, we argue for increasing reflexivity on the contextuality of the production of theories of spatiality and social relations. Thus the world is not only aboard one odyssey, and the odysseys in plural should reflect on the diverse routes of knowledge production. To depart from the social construction of space does not guarantee where exactly one might arrive.

## The odysseys of this book

While it is now generally acknowledged that space is not a neutral medium that stands outside the way it is performed and/or conceived, this does not mean that the conception of space is now presented in a uni-vocal voice. Space figures in many different ways in the strata of current social and cultural writings, and tendencies of the social world at the turn of the millennium seep into our conceptualisations in many ways. While we have no intention of claiming comprehensibility on our account, we can say that the different contributions in the book are trying to distil some main issues in current interpretations of space, either through theoretical discussions or through the exploration of selected cases. One major message is however included in the framing of the book. While it is an imperative to reinterpret our spatial conceptions in the light of current discussions on issues such as (post)modernity, globalisation, mobility and hybridity, the regimes of truth under establishment on the basis on these processes also need qualification. The book discusses the ambivalences, the ambiguities and contextual variations that complicate any attempts to universalise these ideas. Within this background, we can identify some perspectives that connect and cut across the various contributions.

The first issue concerns the role of a supposedly increasing *mobility* in the production of space. It aims to consider how contemporary discourses on mobility influence our conceptions and imaginations of space, and it reaches from questions on experience and identity, over travelling of different kinds, to general accounts of a 'speeded up' or 'mobile' world. This account feeds through to, and off, another one, stating that the world is in the midst of a phase of *globalisation*. Spatialities constructed through restless 'spaces of flows' and relationships between the global and the local are issues in many contributions, as are discussions of established discourses or 'truths' of globalisation and its influence on different spheres of social life. The other side (or scale) of these discussions is about *embodiment*. The human and social body stands out in different ways in a few of the contributions. It appears as a duality between the creativity of embodied practices and the discursive and material construction of social bodies, in this way emphasizing the spatiality of the body – but also in terms of the way in which discourses on 'other' bodies work as an important dimension of the power of geographical imaginations.

These issues in different ways, directly or indirectly, link up with *modernity*. The history of modernity has always been characterized by a movement between dualisms such as order/disorder or fixity/change. On the one hand processes of rationalisation and institutionalisation have been involved in spatial fixation and construction of nation states. On the other one – as Berman so brilliantly illustrated in *All That Is Solid Melts into Air* (1982) – characters of instability, changeability and social emancipation have always been inherent others of these processes. The contributions to the book, in the search for old and new spatialities of modernity, emphasize different aspects of modernity, separately or in combination. We are dealing with the construction of different versions of modernity relating to specific contexts and/or exploring specific themes. The theme of modernity and order is used to emphasize the techniques of *power* and the fixing dimension of modernity. From that perspective, modernity and modern politics are about ordering, about reducing ambivalence and eliminating everything that cannot be precisely defined. Territoriality and the exclusion of 'the other' are important parts of this figure. Another theme, on modernity and mobility, serves as a current counter-image to that one. It emphasizes the disembedding mechanisms of modernity and the globalisation or disintegration of its pre-established institutional order. 'Global flows', 'global networks' and 'non-places of modernity' are some of the notions catching up on the content of this theme, but it also initiates analyses of the de- and/or re-construction of nation states and territorial borders in the light of these processes. Finally, the theme of modernity and reflexivity moves the attention from 'outside' to 'inside' dimensions of modern changeability. It stresses the role of uncertainty and knowledge in social life on the scale of both institutional entities and embodied identities, including 'spatial' identities as reflexive identities.

Reflexivity, in continuation of that, is also part of *geographical representations*. Imaginative geographies in the form of spatial narratives and discourses are part of several contributions, but some of them also connect these issues to production of knowledge itself. That is, they consider the production of imaginative geographies of 'the others' by 'our' societies or 'our' disciplines. The issue at stake in these contributions is the relationship between power, knowledge and space and the importance of turning reflexivity 'inwards' to our own production of geographical knowledge. One topic is the working of geographical representations and spatial stereotypes in a historical perspective. This is a crucial approach to document the socially constitutive powers and practices involved in discourses about the other. Such approaches may help qualifying our reflections on the role of our own contemporary production of knowledge. Another topic is the contextuality and the partiality of knowledge production and power relations involved in contemporary representations of spaces and places. It relates to current discussions on the hegemony of English and North American discourses in 'international' writing spaces (see e.g. Gregson et al. 2003) and underlines the necessary interplay in production of knowledge between transnational communication and contextual knowledges. This reflects what we could see as an additional aim of the book, that is, through inclusion of authors from several European countries to ensure the presence of 'other' voices in the theoretical debate. This strategy also adds to the range of approaches contributing to the common theme. Categorisation of such a group of contributions is, of course, always a risky business, in danger of homogenising the heterogeneous and cutting the connections. However, for the purpose of presentation we have organised the chapters in what we see as three different, but connected, intellectual odysseys through the ground of spatiality and social relations.

## Mobility, immobility and embodied narratives (Part I)

The first journey takes off with John Urry's theoretical elaborations on the irreversible developments of mobile connections. Urry focuses on mobile material worlds and reflects on contemporary social trends by means of theoretical inspirations from, among others, complexity theory (see also Urry, 2003). Mobility and complexity imply the production of systematic instability in a world, where networks are connecting objects, persons and technologies, and there are unpredictable fluids of travellers, migrants, and mobile objects. Power is mobile, performed and unbounded. Citizenships of settled people tend to erode with powers disappearing like sand, and with media-produced scandals trust may disappear overnight. Scandals are 'unpredictable and irreversible' patterns of path dependence, where the metaphor of complexity helps us transcend the distinctions between structure and agency. This first journey challenges conventional understandings of key concepts such as space, society and power.

Travelling and mobility continues into Jörg Bechmann's contribution on the ambivalence involved in mobility versus immobility. Modernity involves the attempt of minimising ambivalence, while all these attempts are in fact only further propelling ambivalences. Inspired by Zygmunt Bauman, Bechmann states, 'the eradication of ambivalence under the modern order, re-solidifies what it seeks to dissolve'. This approach facilitates a study of how mobility is ordered in the case of Copenhagen Airport. It is documented how there are technologies, procedures and practices aiming at producing stability in the fluidity of the airport, where travellers are in a permanent state of ambivalence of whether or not to travel, to fly or not to fly.

The contributions of the following two female geographers are somehow more sceptical to the notions of mobility developed in the first two contributions by male sociologists. Neither the general difference nor the gendered or disciplinary character of the division should be exaggerated. However, although in different ways, both Kirsten Simonsen and Anke Strüver emphasise the embodied and narrative character of spatial practices, and for both the contribution of Michel de Certeau is part of the inspiration.

The multiple faces of cities form the point of departure in Kirsten Simonsen's contribution. Simonsen is sceptical to discourses on increased mobility in general, and she suggests looking further into the power-geometries of social inequality and gender in respect to mobility. Furthermore she challenges the possible exaggeration of technology-generated mobility, as well as the univocal trends in some discourses on mobility. 'Rhythm-analysis', as it is formulated in the late works of Henri Lefebvre, inspires the development of a framework for studying the multiple spatialities and temporalities involved in the construction of the city. The contribution develops a theoretical approach to the combined understanding of embodiment and the narratives of the city, where narratives are more than just memories. Narratives also anticipate by their capabilities of spatialising actions in multiple ways.

This first part ends with the 'space oddity' thought experiment in Anke Strüver's contribution, where the disappearing space traveller loses contact with earth in a space beyond ground control. This is a metaphor for the European Union's visual simulacrum about intensive cross-border mobility that Anke Strüver considers doomed to misunderstand the everyday practices of people along the border between

Germany and the Netherlands. Her analysis illuminates the spatial practice effects of cognitive-imaginative borders since these are firmly linked with the narratives of everyday life. The contribution discusses how, theoretically and empirically, to transcend the dualism of representations and practices. It is a study of 'performed immobility' in which groundless invitations to cross borders are confronted with the reality of imaginative borders.

This first odyssey, then, raises important questions about the consequences of mobility and immobility. It does so in ways that also reflect on how we produce ideas, visions and knowledge about mobility. All four contributions do so in ways that are committed to thinking about how humans do not lose 'ground control'.

## Territoriality, mobility and identity politics (Part II)

Our second odyssey passes through various versions of the construction of modern orders, exploring how practices and identities of these are both governed by principles of mobility and territoriality. The five contributions take us on a journey from the spatial policy discourse of the European Union, over the territorial principles of the nation-state symbolised by passports and local cases of socio-spatial construction in the peripheries of Iceland and Greenland and the unfinished state construction in the Andean societies in Peru, culminating in a discussion on globalisation and everyday life. The travel is navigating in-between territoriality and mobility.

Our second odyssey then starts in the world of the discourse of European spatial planning. Ole B. Jensen and Tim Richardson take us into the realm of frictionless society, bringing us further along the lane of mobility. They propose a practice- and culture-oriented approach to spatiality and mobility that can unfold the dialectics of socio-spatial practices and symbolic-cultural meanings, while also adding the politics of scale. They see modernity as mobility and immobility intertwined. As such they recall central elements of the discussion in the first part of the book. The vision of a Europe without constraints, where the pursuit of growth tends to overshadow those of ecology and equity, is a planning discourse that carries the will to order. Much of the energy comes from second order governance of lowly transparent networks outside of democratic control. It produces the hegemony of a Europe of flows that seems to have become embedded. Paradoxically, this discourse is in fact aiming at producing territorial identification with Europe.

The territorial order of the nation-state does however seem to be more solid than many think, and this is the point made in Wolfgang Zierhofer's contribution. He considers the nation-state a central mediator and translator in between world-society and individual subjects, and the nation-states constituted in interaction with both of them. The cosmopolitan principle inherent in the institution of the passport is a significant part of this dominant order, where the differentiation between micro and macro is transcended. Still, the modern order produces spatial scales as social spaces that localise activities and produce identities and 'communities of common destiny'. Nation-states are active co-producers not only of the globalisation of capital but also of cosmopolitan awareness of the world-society; they are both part of and constitutive to the development of the inter-state system. However, there are new trends to overcome those binary oppositions that constitute the political in modernity and which seem to facilitate non-modern political spheres, where so-called sub-politics is no longer sub-ordinary but politics like all other.

Our journey now takes us to the Nordic Atlantic peripheries where not only nation-states, but also colonial states and local municipalities, have been producers of significant modern orders. Jørgen Ole Bærenholdt's contribution is about the traditions of territorial entrepreneurship and empowerment of inhabitants in costal localities. Coping is the 'way people engage in strategies that makes sense to themselves', and these are creative but yet also modest socio-spatial practices of survival. Coping combines not only the spatial practices of territoriality and mobility, but also the social practices of bridging and bonding in diverse ways of innovation, networking and formation of identity. The cases of local development in Iceland and Greenland point to the increasing influence of mobile business in fisheries and tourism; this to an extent challenges the territorial bridging of local citizens across the social distinctions involved in international business development.

Next, we continue to the Andean societies in Peru, where Fiona Wilson contextualises mobility in a 'vertical system' of resource control in the Andean mountains. In contrast to the three preceding contributions, this is a context where modern states were never really established. Therefore road construction projects are becoming significant cases of the meaning of mobility. This is especially the case because various territorial regimes have been coexisting as overlapping spaces, including those of peasant and indigenous response to the attempts of central powers. We are brought into the conditions of civil war, where spaces of the state, haciendas, indigenous people and the Sendero strategy of isolation are in conflict. Thus, the construction of a road becomes a symbol of citizenship of the absent state arriving. This is a situation of identity politics played out in plural and in conflict.

The last chapter in this section in some sense explores a different road. Like the other authors, Benno Werlen is interested in processes of globalisation, territorialisation and regionalisation, but from a perspective of everyday life. In order to pursue that goal, he seeks an ontological grounding of the discussion and attempts to develop an ontology of socio-spatial relations based on subjective action. Part of that is a re-vitalisation of Husserl's concept of life-world. From that, he sees regionalisation in a globalised world as a practice of 'world-binding', that is, a practice of re-embedding that brings the 'world' within the reach of actors by way of their allocative, authoritative and symbolic resources.

## The spatial production of knowledge (Part III)

Geographical imaginations and spatial stereotypes are produced in specific spatial and temporal orders, and this is also truly the case with this book. In order to illuminate this point, our third and final odyssey is of another kind. It brings us into the histories and geographies of the production of knowledge.

Costis Hadjimichalis and Dina Vaiou directly address the contextuality of development of geographical theory. From their position in South European Greece they bring us two critical cases of contested geographical imaginations. The first one addresses the concept of the informal, reflecting on its ambiguous position in contemporary social and geographical theory. It is either left out, they state, or it is given the position of the less developed other, that is, when informal economies are acknowledged it is in the role of being deviant from the 'norm'. Hence, dominant discourses make sense of the informal in terms of flexible economies and entrepreneurship that does not acknowledge the specific social and cultural relations involved in these practices in 'the South'. The second case is about political and cultural geography discourses of the Balkans. Ironically, while celebrating borderless worlds and the other, even progressive geographers expressed discursive support of NATO military intervention. Here, the image of '*Balkanism*' travels in ways that do not reflect on the origins of this naming. Hence, the Balkans becomes a fixed notion of the political unstable other, a discursive strategy that does not leave much space for other voices to be heard.

Keld Buciek's travel takes as a point of departure social control strategies and the power of classification. The argument is that modernity could hardly succeed without establishing an idea of the unhealthy other, an atavistic being, by whose sole existence society risks degeneration. Having been constructed during the nineteenth century, this idea has shown a remarkable persistency in relations between knowledge, geography and 'socio-medico' labels classifying those others whom society regards as potentially dangerous. Consequently the idea of purifying has been a dominant metaphor to explain social practices at home as well as abroad. It is argued that inclusion and exclusion of 'the other' at home and abroad through means of labelling the other 'in-pure' or 'unhealthy' still are parts of the ordering processes of modernity, not at least in immigration policies.

On the same line, Derek Gregory in the final chapter of this section – and of the book – takes us on a spatio-temporal journey irrefutably illustrating the power and the spatialisations of imaginative geographies. Starting from Edward Said's critique of Orientalism he shows how 'colonial modernity' produces its other in inherently asymmetric ways. But he soon begins to argue that this cultural colonialism is not merely a matter of history, it reaches forward into the formation of our own 'colonial present' (see also Gregory 2004). Through considering some of the dichotomic and parochial tropes circumscribing the post-September 11 debates, Derek Gregory challenges some of the current 'truths' of global spatialisations. The power to compress distance, he argues, is also the power to expand distance. So, if global capitalism is aggressively de-territorialising, colonial modernity on the other hand is intrinsically territorialising, continually distancing 'us' from 'them'. Current imaginative geographies are thus doubled spaces of articulation, at the same time establishing connections and separations.

**Modalities of space**

By way of a conclusion, we will try to draw out some of the ways in which space figures in the odyssey of the contributions to this book. Using the notion of modalities of space we want to indicate a two-step journey. First, all contributions in one sense or the other move away from the Kantian perspective on space (as an absolute category) towards space *as* process and *in* process – that is, space combined with time. Secondly, and as a consequence of that initial concurrence, the emphasis moves from the general discussion of social spatiality towards explorations of more specific modes of space. The modalities and the ways in which they figure (partly overlapping) in the different contributions reflect 'concrete' developments in contemporary social configurations as well as different philosophical/theoretical positions.

The first mode of space that attracts attention is exactly one of those more 'concrete' ways of thinking spaces of modernity. With reference to the work of Castells (1996) we have chosen to call it *spaces of flows*, even though it is taken in through a broader set of narratives of 'time-space compression' circulated by authors such as Castells, Harvey (1989) and Bauman (2000) alike. The historical base of this conception dates back to the nineteenth century where the gathering speed of means of communication began to enable people to travel and communicate more swiftly, thus bringing places closer together in time and rewriting the horizons of experience, including notions of space. As this process is continued, it is argued, the contemporary world reached a new phase of 'globalisation'. The process of speed-up, particularly boosted by new electronic communications media, reaches a new plane producing restless spaces of flows and mobile patterning of social life. Spaces are trajectories, thought of in terms such as fluidity *vs*. viscosity, harmony *vs*. discordance, velocity and turbulence. And the flows consist of money, commodities, technology, information, imaginations and people. The last-mentioned, travelling people, intermittently encounter each other in what Augé (1995) called the 'non-places of modernity' (airport lounges, coach stations etc).

This story about globalisation, speed-up and spaces of flows has (for good reason) become increasingly commonplace in the literature; it is often more or less taken for given in both public imagination and academic writing. Also in this collection, the story of spaces of flows in many contributions constitutes the silent partner from which other aspects of current spatialities are explored. It is however also directly addressed. It is developed and qualified by way of new theoretical inputs, such as 'complexity theory' and the language of the 'new physics'. But it is also challenged in different ways; by including analysis of 'politics of scale', by addressing the alleged one-directed and univocal understanding of time-space involved, and by unearthing the contradictory logics of time-space compression and time-space expansion in contemporary global constellations of power.

This can lead to our second modality of space, which we tentatively have called *spaces of boundaries*, and which in the current collection most of all refer to the relationship between globalisation and territoriality. One important point from the discussions is the incitement to consider this relationship, not as one of contradiction, but as one of connection. For instance, nation-states are themselves part of and presupposing an international order, a universal code of nationhood. They are, in the words of Wolfgang Zierhofer, both 'global localisers' and 'local globalisers'. Imposition (and transcendence) of boundaries is involved in both enactment of power and construction of identity, and it is in the contributions explored on scales reaching from local communities to the European Union. And, as discussions on mobility and territoriality suggest, territories should not be seen as bounded spaces only. Whilst some boundaries definitely have hardened over (notably around migration controls in Europe), others are less impermeable and are transcended on a daily basis. Boundaries and bridges in a complex sense are two sides of the same coin. 'Open walls' is a metaphor suggested by Allen (2003) to signify such porosity of boundaries.

The notion of daily movements brings us, in a different incarnation, to a third way of thinking space, through the *spaces of experience*. Immediately such a notion, represented especially but not only through phenomenology, may seem problematic because of its traditional attachment to a self-centred subject. However, both in parts of phenomenology itself and other streams of philosophy and social theory, this connection is efficiently loosened. They all in some way or other relate to issues of *practice*. One such stream concerns social conceptions of the body and *embodiment*. Embodied (sexed and gendered) practices are spatial and they spatialise – at the same time incorporating structures and norms from the environment and creating it through realisation and connectivity. Perhaps the encounters of such moving bodies, with their variety of spatialities and temporalities, are most illustratively brought to life by Lefebvre in his tentative notion of 'rhythmanalysis'. Everyday practices and moving bodies further draws attention to the issue of *mobility*, which is a common issue in many parts of this book. Writing from experience today increasingly involves travel and mobile assemblies of space and time – for example, as identified by Jörg Bechmann, the indelible moments of 'ambivalence' involved in modern (mobile) ways of being in the world.

Practising and thinking about space also occur through the medium of language – opening up towards *spaces of imaginations*. In the first place, and closest to theories of practice, this can refer to *narratives* as something we live by, as the medium through which we order and make sense of experiences and events – personal as well as social. Such narratives are temporally and spatially constitutive and thus part of everyday constructions of space by moving bodies (here represented in relation to the construction of urban environments and to the re-constitution of borders in a mobile world). A more representational line of thought adds issues of power and *discourse*, as in the analysis of construction and effects of spatial policy discourses. And further along the line of power come works based on Said's critical triangulation of power, knowledge and geography in his 'imaginative geographies' of the Orient. They give rise to explorations of the relationship between modernity and colonial discourse and of spatialities of the colonial present. But they also generate discussions of different processes of othering in different scales – as for instance the different processes at work in the construction of 'Orientalism' and 'Balkanism'.

Finally, we can close the circle by identifying a modality of space that departs from both social experiences and imaginations, but at the same time links to spaces of flows. This is what we call *spaces of the transhuman*. They originate from an increasing attention to (the active role of) the object world, such as the computer Hal in *2001: A Space Odyssey*. Taking inspiration from 'actor-network' theory and related approaches, they refer to various hybrids consisting of bodies, texts, machines, architecture etc. and possessing the power of rapid movement over many regions. The spatialities in question are relational spaces of connectivity; global networks or fluids connecting people, objects and technologies across multiple and distant spaces and times. The above-mentioned translation from 'new science' or 'complexity theory' is, in particular, directed towards description of this modality of space.

These five and most likely more modalities of space intersect in various unexpected ways. The ambivalence of modernity has certainly not been less since September 11, 2001. The world seems to become ironic in a 'dark' sense. *Spaces of flows* were supposed to bring about the speed-up of connections between people, but became the means of separation of people. Likewise, we used to think of *spaces of boundaries* as processes of separation and exclusion, while boundaries in many cases reveal themselves as spaces of connection and interaction of people, states, social movements and businesses. Much evidence for these double-bound processes are found in the *spaces of experience* where the embodied practices of people's encounters with other people and objects involve coping with the often unpredictable, unreliable, switches between separation and connection. People's mobile practices also occur in *spaces of imagination*, in the narratives people live with and through. But the narratives of everyday life is bound up with the more or less global discourses of spatial policy, cross-border interaction, transport development, geopolitics and empire. Various forms of banal globalism, nationalism and colonialism intertwine in everyday practices, thus also feeding and supporting the more explicit and violent forms of globalism, nationalism and colonialism, that often work more in concert, that we anticipated.

Again ironically, flows become the most effective performance of the imagined 'clash of civilisations'. Truly the play of the clash has been scripted in advance, but the energy of making the performance also emerges from the *spaces of the transhuman*. Of course, there are still, it seems, such Power containers that instruct the play to begin, and could, if they possibly would, also let it come to a stop. Besides the clashes of powerful states and global networks about control, resources and identities, this could also be true of global environmental and social problems. So the interest for the role of non-human objects and flows should never be accepted as a bad excuse for humans for not taking responsibility for the actions, encounters, narratives, separations, connections, violence and possible humility they actually have some powers to limit, control, cope with and perform. The plea for scientific approaches to space that overcome binary conceptualisations put forward in this book, also imply a claim for recognition of human reciprocity.

## References

Allen, J. 2003 *Lost Geographies of Power*, Oxford: Blackwell.
Augé, M. 1995 *Non-places*, London: Verso.
Auriac, F. and Brunet, R. 1986 (eds) *Espaces, jeux et enjeux*, Fayard: Fondations Diderot.
Berman, M. 1982 *All That Is Solid Melts into Air*, New York: Simon and Schuster.
Bauman, Z. 2000 *Liquid Modernity*, Cambridge: Polity.
Castells, M. 1996 *The Rise of the Network Society*, Oxford: Blackwell.
Gregory, D. 1994 *Geographical Imaginations*, Oxford: Blackwell.
Gregory, D. 2004 *The Colonial Present*, Oxford: Blackwell.
Gregory, D. and Urry, J. 1985 (eds) *Social Relations and Spatial Structures*, London: Macmillan.
Gregson, N., Simonsen, K. and Vaiou, D. 2003 'Writing (across) Europe: On writing spaces and writing practices' *European Urban and Regional Studies*, 10(1): 5–23.
Harvey, D. 1989 *The Condition of Postmodernity*, Oxford: Blackwell.
Lefebvre, H. 1991 (orig. 1974) *The Production of Space*, Oxford: Blackwell.
Tonboe, J.C. 1985 (ed) *Farvel til byen?* Aalborg: Aalborg Universitetsforlag.
Urry, J. 2003 *Global Complexity*, Cambridge: Polity.

# PART I
# MOBILITY, IMMOBILITY AND EMBODIED NARRATIVES

# Chapter 1

# The Complex Spaces of Scandal[1]

## John Urry

### Global processes

This chapter considers whether complexity can generate productive metaphors for the analysis of various 'post-societal' worlds. I follow complexity-theorist Brian Arthur's views that what complexity writers are doing is: 'beginning to develop metaphors' and that the Santa Fe Institute, for example: 'is in the business of formulating the metaphors for this new science, metaphors that, with luck, will guide the way these sciences are done over the next fifty years of so' (1994: 680). I consider whether complexity can provide some metaphors relevant for thinking through the evolving nature of spatial relationships in the twenty-first century (for more details, see Urry 2003).

Special focus will be placed upon the metaphors appropriate for examining the fused 'material worlds' implicated in the apparent 'globalisation' of economic, social, political, cultural and environmental relationships. In the past decade the social science of the global has extensively described many of these relationships. However, it has not developed analysis of the emergent properties at the system level. It has tended to take the global for granted and then shown, how and in what ways, various localities, regions, nation-states, environments and cultures have been transformed in linear fashion by this all-powerful entity called 'globalisation'. Thus globalisation (or sometimes global capitalism) has come to be viewed as the new 'structure', while nations, localities, regions and so on, the new 'agent', to employ conventional social science distinctions but given a global twist.

This chapter uses complexity to try to move beyond various positions within social theory. Complexity maintains that there is no 'structure' and no 'agency', no 'macro' and no 'micro' levels, no 'societies' and no 'individuals', and no 'system-world' and no 'life-world' – in that each of these is presumed to be separate and distinct essences brought into external juxtaposition with its other. Overall the argument here is one of 'relationality', a position not only central to complexity but also to actor-network theory and various post-structuralist formulations. Dillon characterises relationality as being where: 'No party to a relation is therefore a monadic, or molar, entity. Each is, instead, a mutable function of the character of the mode-of-being-related and its capacity for relationality' (2000: 12). And relationality is effected through a wide array of networked or circulating relationships implicated within different overlapping and increasingly convergent *mobile, material* worlds.

The linear metaphor of scales, such as that stretching from the micro level to the macro level, or from the lifeworld to system world, that has plagued social theory from its inception, should thus be replaced by the metaphor of connections. Such connections are to be viewed as more or less mobile, more or less intense, more or less social, and more or less 'at a distance' (see Dicken, Kelly, Olds, Yeung 2001: 102–4; Sayer 2000, on system/lifeworlds). Latour maintains that the social 'possesses the bizarre property of not being made of agency or structure at all, but rather of being a *circulating* entity' (1999: 17). There are many trajectories or mobilities that are neither macro nor micro but circulate between each in terms of 'speed; velocity; waves; continuous flow; pulsing; fluidity and viscosity; rhythm; harmony; discordance; and turbulence' (Dillon 2000: 12). There is therefore no top or bottom of society, but many connections or circulations that effect relationality through performances at multiple and varied distances, as I examine in the case of what we can call 'global scandals'. Latour summarises how: 'there is no zoom going from macro structure to micro interactions ... [since] both micro and macro are local effects of hooking up to circulating entities' (1999: 19).

**Complexity theory and time-space**

This chapter is therefore concerned with the systemic non-linear relationships of global complexity that transcend most conventional divides of social science (on complexity, see *inter alia* Waldrop 1994; Prigogine 1997; Capra 1996, 2002; Byrne 1998; Cilliers 1998; Urry 2003). First, the very large number of elements makes such systems unpredictable and lacking any finalised 'order'. Such elements interact physically and, because of de-materialising transformations, informationally over multiple time-spaces. They are irreversibly drawn towards various 'attractors' that exercise a kind of gravity-effect. Interactions are complex, rich and non-linear involving multiple negative and, more significantly, positive feedback loops with ineluctable patterns of increasing returns and path-dependence stretching over time-space. Systems interact dissipatively with their environment. The elements within any such system operate under conditions that are far from equilibrium, partly because each element only responds to 'local' sources of information. But elements at one location have significant time-space distanciated effects elsewhere through multiple connections and mobile trajectories. There can be a profound disproportionality of 'causes' and 'effects'. Systems possess a history which irreversibly evolves and where past events are thus never 'forgotten'. Points of bifurcation may be reached when the system branches. Such systems should never be seen as involving *simply* linear increases in the colonisation of the life-world, or of enhanced agency, or of greater risk, or the domination of the capitalist market.

Such complexity notions problematise 'social order'. In the cybernetically-influenced writings of Talcott Parsons, for example, there is a hierarchy of values and norms that works through each society at all levels, a clear notion of social equilibrium, and strong negative feedback or steering mechanisms that can rapidly and effectively restore order. But the implications of complexity is that there never is such a clear and effective set of processes and indeed that efforts to restore order always engender unforeseen consequences, often of a kind that may take the society further away from any ordered equilibrium. Moreover, Parson and the classical tradition little considered the mobile patterning of social life which problematises the fixed, given and static notions of social order. Ordering, one might say, is always achieved 'on the move'.

Moreover, social order is not the outcome of uniquely social processes. As Law argues: 'the notion that social ordering is, indeed simply social also disappears. ... what we call the social is materially heterogeneous: talk, bodies, texts, machines, architectures, all of these and many more are implicated in and perform the social' (1994: 2). In that sense sociology's notion of accounting for a purified *social* order is no more and should be relegated to the dustbin of history (see Latour 1993; Knorr-Cetina 1997).

Thus criss-crossing 'societies' are diverse systems in complex interconnections with their environments, there are many chaotic effects time-space distanciated from where they originate, there are positive as well as negative feedback mechanisms that mean that order and chaos are always intertwined, there are self-organising global networks and global fluids moving systems far from equilibrium, and there is not social order accounted for by purified social processes. Such complexity-thinking enables the transcendence of the dichotomies of determinism and free will, especially through seeing material worlds as unpredictable, unstable, sensitive to initial conditions, irreversible and rarely 'societally' organised.

Especially important here are unpredictable and yet irreversible patterns of path dependence, such patterns emphasising the importance of the order of events or processes, as we see below in the case of contemporary scandals. *Contra* linear models, the order in time and across space very significantly influences the way that processes turn out (Mahoney 2000: 536). Causation can flow from a contingent minor event to general processes that demonstrate great significance. Such path dependence can be established for particular small-scale, localised reasons. Two well-known examples are the QWERTY keyboard introduced in 1873 to *slow down* typists so that the keys did not jam, and the dominance of the petroleum-based automobile over preferable fuel alternatives developed in the 1890s before 'path-dependence' of the petroleum-based car had set in. What then is key is that: 'small chance events become magnified by positive feedback' and this 'locks in' such systems so that massive increasing returns or positive feedback result over time (Brian Arthur, quoted in Waldrop 1994: 49; Mahoney 2000). There are relatively deterministic patterns of inertia that reinforce established patterns through positive feedback. This escalates change over time away from what we might imagine to be the point of 'equilibrium' and from what might be optimal in 'efficiency' terms, such as a non-QWERTY keyboard or non-petroleum based forms of automobile fuel (but see Mahoney 2000, on sociological rather than utilitarian path dependence models).

Positive feedbacks and path dependence, where contingent events can set in motion institutional patterns with deterministic outcomes, are central to the power of various global networks (and to other sets of events such as global scandal). These networks are large in size, may involve dense interactions within their nodes and may interact with other networks so expanding further their exceptional range and influence. These networks consist of various hybrids that roam the globe, possessing the power of rapid movement, across, over and under many apparent regions (Bauman 2000; Urry 2004). Elsewhere I show that there are two main forms of such hybrids (Urry 2003). First, there are global networks like that characterising McDonalds with a tightly coupled network consisting of complex, enduring and predictable connections between peoples, objects and technologies across multiple and distant spaces and times (Murdoch 1995: 745; Law 1994: 24). Relative distance is a function of the relations between the components comprising that network. The invariant outcome of a network is delivered across its entirety in ways that often overcome regional boundaries. Things are made *close* through these networked relations. Such a network of technologies, skills, texts and brands, a global hybrid, ensures that the same 'service' or 'product' is delivered in more or less the same way across the entire network. Such products are predictable, calculable, routinised and standardised. Many 'global' enterprises organised through such networked relations, such as McDonalds, American Express, Coca Cola, Microsoft, Sony, Greenpeace, Manchester United and so on (see Ritzer 1998; Klein 2000).

Second, there are various global *fluids*, such as world money, automobility, digitised information, the internet, social movements, international terrorism, travelling peoples and so on. Global fluids travel along various scapes but they may escape, rather like white blood corpuscles, through the 'wall' into surrounding matter and effect unpredictable consequences upon that matter. Fluids move according to certain novel shapes and temporalities as they break free from the linear, clock-time of existing socio-scapes – but they cannot go back spatio-temporally because of irreversibility. Such fluids result from people acting upon the basis of local information but where these local actions are, through countless iteration, captured, moved, represented, marketed and generalised, often impacting upon hugely distant places and peoples. Such fluids demonstrate no clear point of departure, just de-territorialised movement, at certain speeds and at different levels of viscosity with no necessary end-state or purpose. This means that such fluids create over time their own context for action rather than being seen as 'caused' by such a context. One such global fluid is the internet, in a way invented in 1990 and which has developed into an irreversible autopoeitic system, especially following the 'chance' invention of the first web browser in 1993/4. Plant argues that: 'No central hub or command structure has constructed it ... This was one of the first systems to present itself as a multiplicitous, bottom-up, piecemeal, self-organizing network which ... could be seen to be emerging without any centralized control' (1997: 49). The Web possesses a non-hierarchical rhizomatic structure based upon lateral, horizontal *hypertext* links that render the boundaries between objects as fluid.

Thus global complexity is comprised of many different 'islands of order' within a sea of disorder. There are global networks and global fluids; and there are also national societies, diasporas, 'supra-national states', global religions or 'civilisations', international organisations, international meetings, NGOs and cross-border regions (Habermas 2001: chap 4). And a single state thus finds diverse self-organising networks, fluids and 'polities' seeking to striate its space. States have shifted away from governing a relatively fixed and clearcut national population resident within its territory and constituting a clear and relatively unchanging community of fate, what I used to call 'organised capitalism' (Urry 2000: chap 8). Shifts towards global networks and fluids transform the space beyond each state that they have to striate, such as in the case of scandal. Habermas argues ' "globalization" conjures up images of overflowing rivers, washing away all the frontier checkpoints and controls, and ultimately the bulwark of the nation itself' (2001: 67). States thus act as a legal, economic and social regulator, or *gamekeeper*, of practices and mobilities that are predominantly provided by, or generated through, the often unpredictable consequences of many other entities. Social regulation is both necessitated by, and is made possible through, new computer-based forms of information gathering, retrieval and dissemination (Power 1994).

Thus the fluid and turbulent nature of the global complexity means that: 'the role of the state is actually becoming more, rather than less, important in developing the productive powers of territory and in producing new spatial configurations' as with the US-led global coalition against terrorism (Swyngedouw 1992: 431). Indeed there has been an 'enormous expansion of nation-state structures, bureaucracies, agenda, revenues and regulatory capacities since World War II', in order to deal with multiple and overlapping global fluids that move across borders through time-space in dizzying, discrepant and transmutating form (such as from students to tourists to terrorists). States are not converging in a uniform powerless direction but are becoming more diverse, such as the US, the EU and the Taliban (Weiss 1998: chap 7).

## Power and complexity

Much thinking about power in the social sciences has been focused upon the interrelationships between agents or subjects. Power is conceptualised as an attribute of agents, through observing two or more human agents and seeing in what ways, and to what degree, the actions of each are influenced by that of the other. Lukes' *Power. A Radical View* famously critiques this inter-subjective conception of power through the advocacy of a three-dimensional view organised around the notion of 'real interests' (1973). Lukes shows how the most effective exercise of power occurs when there is no overt or even covert bending of the will of one agent by that of the other. Power is exercised if people's real interests are secured and this is best realised without overt or covert inter-subjective competition and struggle. But in much social science power conceived as a property of agents remains central to the analysis of social relations.

However, complexity transcends the division between free will and determinism and hence between agency and structure. It transcends the characteristic way in which power has been located, as agency. So what then would constitute a complexity approach to power? Power would not be regarded as a thing or a possession. It is something that flows or runs and it has become increasingly detached from specific territory or space and especially non-contiguous. Bauman outlines a 'post-panoptical' conception of power (2000: 10–14). Power is not necessarily exercised through real co-presence as one agent gets another to do what they would otherwise not have done through interpersonal threat, force or persuasion. But also power no longer necessarily involves the imagined co-presence of 'others' within a literal or simulated panopticon. Rather the prime technique now of power is that of 'escape, slippage, elision and avoidance', the 'end of the era of mutual engagement' (Bauman 2000: 11). Modern societies had involved a mixture of citizenship with settlement and hence with co-presence within the confines of a specific territorially-based society. But now the new global elite, according to Bauman, can rule: 'without burdening itself with the chores of administration, management, welfare concerns', even involving developing disposable slave-owning without commitment (2000: 13; see Bales 1999, on 'disposable peoples'). Travelling light is the new asset of power. Power is all about speed, lightness, distance, the weightless, the global, and this is true both of elites and of those resisting elites, such as anti-globalisation protestors or bio-terrorists or those seeking to reveal a scandal. Power runs in and especially jumps across different global networks and fluids. Foucault famously described the shift from sovereign power to disciplinary power (1977). But now there are further shifts towards *informational and mediated* power. Citizenship and social order has always depended upon relations of mutual *visibility* between the citizen and the state. But by the twenty-first century citizens are subject to informational mediated power, forms of power that are complex in their mechanisms and consequences.

First, such power is enormously technologised with the development of vision machines, tens of thousands of satellites, bugs, listening devices, microscopic cameras, CCTV, the internet and new computerised means of sharing information (see Lyon, 2001, on the post September 11th surveillance effects).

Second, every day life also increasingly involves speed, lightness, and distance, with the capacity to move information, images, and bodies relatively unnoticed through extensively surveilled societies. Resistance to power is also mediated and highly fluid-like.

Third, such mediated power functions like an attractor. Within the range of possibilities, the trajectories of systems are drawn to 'attractors' that exert a gravity effect upon those relations that come within its ambit. The global media exert such a gravity effect, with almost the whole world both 'watching' and being seduced into being 'watched' (as with the videos of Bin Laden). The attractor is rather like the game show 'Big Brother' written onto the global screen, as we see in the next section on scandal.

And fourth, power is mobile, performed and unbounded. This is its strength and its vulnerability. Attempted ordering by the most powerful (threatened by scandal) can result in complex unintended effects that take the system even away from equilibrium. In such unpredictable and irreversible transformations, mediated power is like sand that may stay resolutely in place forming clear and bounded shapes with a distinct spatial topology (waiting say to be arrested or bombed) or it may turn into an avalanche and race away sweeping much else in its wake. And correspondingly, challenging that power is also so hard since bombing certain nodes of power cannot destroy the 'lines of flight' that simply flow like 'packets' in email systems and following different routings and getting round destroyed nodes. The power of the detested US is not anywhere in particular and so in a sense can never be eliminated, except perhaps by a nuclear winter that destroyed all the informational and media power scapes across the world.

I now examine some aspects of this mobile and complex notion of power through the example of scandal.

**Complex scandals**

The late twentieth century emergence of a 'mediated power' criss-crossing the globe produces distinctly new forms of incredibly mobile, mediated scandals (Thompson 2000). Global media disembed events from local contexts and move them, often instantaneously, simultaneously and irreversibly across the globe. At the same time, the breaking down of more solid class-based form of politics means that relationships are in most countries and regions less organised and more fluid and mobile – more wave-like – and hence scandal events more rapidly emerge, and pass in, through and beyond the frontiers of given societies. Four processes in combination generate 'complexity' outcomes with regard to contemporary scandals; normative transgressions, the vulnerability of trust, the fact of exposure, and the power to make events instantaneously and simultaneously visible across the globe.

First, globally mediated scandals occur where there are significant transgressions of particular *norms* of 'expected behaviour' that characterise a given society or type of society. According to Thompson these transgressions normally relate to sexual behaviour, financial matters or the use/abuse of power. Given the ambivalent, contested and often quite strict norms relating to public figures and institutions, then 'scandalous' transgressions do regularly occur (as most citizens are well aware!). There are many potential scandals especially since public figures and institutions are normally confronted by stricter norms of what is appropriate behaviour, as opposed to those not in the mediatised 'public' eye. But the public eye seduces increasing numbers of new 'subjects' that are drawn into the visual media and who are then be subject to cycles of transgression, revelation and confession (what we might term Big Brother narcissism).

Second, the power of certain incumbents of official positions, of companies and of states rests upon a 'politics of *trust*'. In the case of particular incumbents such trust is often based on their presumed character rather than on specific skills. Sometimes there appears to be a kind of 'global trust' (such as that on occasions enjoyed by the President of the US, the General-Secretary of the UN, Nelson Mandela, certain states, certain global companies and so on). But such trust has to be continuously earned or performed and hence it can rapidly erode. There is much to lose especially where trust and character is core to the establishment and maintenance of the legitimacy of an incumbent or organisation. And the greater the trust, the more that character has been put on line, then the greater is any ensuing scandal (other things being equal). Trust is an exceptionally strong but an incredibly brittle resource. It has to be continuously performed. If it stops being earned then it will erode instantaneously as with an individual whose character or good name gets exposed and subject to scandal. Trust may disappear overnight rather like sand can suddenly slip through one's fingers (see Cilliers 1998, on piles of sand). As those subject to scandal often say, although they took years building up their good 'name', 'their world collapsed overnight' as the scandal 'swept' over both them and their unfortunate friends and family.

Third, there is the power of *exposure*, the making of the private transgressive act transparent to the public and hence just that, public (see Balkin 1999; Meyerowitz 1985). The wide-ranging global media increasingly possess the techniques to make transparent what the powerful would seek to maintain as 'private'. Such media employ technologies of observation, surveillance and monitoring of people within their supposedly 'private' lives. These technologies were initially developed within the secret services of states, such as eavesdropping, phone-tapping, secret cameras, listening devices, telephoto lenses, computer hacking, stalking and so on (with Watergate of course it was Nixon's own tapes that provided his downfall). Exposure makes what is supposedly backstage or 'private', frontstage or 'public'. And with digitisation there are few if any images of the private that can ever be 'locked away' for good, which will remain forever private and opaque.

Fourth, there is the attractor of *visibility or tranparency*. With mediatisation power is increasingly visualised. People and organisations across the globe get drawn into the ambit of visibility and its seductive charms to be famous for 15 minutes. And it is bodies that are made especially visible through mediatised visibility, bodies that speak to the world both intimately and 'close-up' and yet simultaneously to vast numbers (as with the Bin Laden videos). This produces a distinct kind of performative bio-power, an embodied power or a 'public intimacy' through figures made visible and revealed on the world's media. But this immense bio-power is exceptionally vulnerable – to exposure as those 'figures' of power can be suddenly, overnight, seen (through) as flawed, as bodily scandalous. All those watching on the media can bear witness to the public shaming, the making transparent, of the transgresser and on occasions their public confession. Gitlin presciently described this as *The Whole World is Watching* (1980). The exposed individual, company or state is revealed, their scandalness is made visible and their bio-power dissolves in front of the world's gaze. Moreover, the competitive nature of the overlapping and digitised media enable the figure of the 'wrongdoer' to be revealed, replayed again and again and their global shame made visible before, and endlessly repeated across, the globe (as paradigmatically with President Clinton and the fascination with his immense but vulnerable bio-power).

Thus scandals are all about how small causes (the furtive embrace, the tiny lie, the small payment, the handwritten note) can, in certain circumstances, produce distant and catastrophic consequences for those involved, and indeed for many others drawn into the swirling vortex of a scandalising event (see Thompson 2000: 75, on the unpredictability of the course of scandal). There are unexpected, unpredictable and uncontrollable visibilities as images flow within, and rapidly jump across, the various media. They compete for global stories and produce what Balkin terms a kind of 'cascade effect' with different journalists with diverse standards of journalistic integrity competing with each other for further scandals to reveal (1999: 402). Especially crucial in producing such cascades are visual images that disrupt or ridicule or overturn existing relations of bio-power. Such images get endlessly sold and resold across the globe, as they subvert, humiliate and transgress the apparent power of the powerful. Those who live by the media can also die through such mediated cascades.

Scandals thus involve complex sets of events that are irreversible *and* unpredictable. They run out-of-control once there is exposure because of the *mobility and speed* of the processes of exposure, visualisation and re-circulation. The irreversible events lead away from equilibrium especially where those involved seek to manage events, to cover their tracks, to lay the ghost to rest. There are powerful positive feedback mechanisms where character and trust dissolve almost overnight and the shame is massively enhanced and magnified. In complexity-mode, most attempts to minimise exposure result in the enlargement of the scandal, and especially the further scandal of 'concealment'. Efforts to dampen down the evolving events produce complex magnification with diverse and further irreversible consequences.

Indeed once the media have peered 'backstage' then there is escalating exposure and visualisation of the scandalous figures. Balkin describes the 'self-amplifying focus' of a media feeding 'frenzy' that takes root and leaves little standing in its whirlwind path (1999: 402). Scandals can possess a kind of all-consuming flow that can 'wash' over those caught up in its wake. As Thompson argues: 'the experience is likely to be overwhelming, as events rapidly spin out of control', away from any movement towards equilibrium (2000: 85).

Such liquid mediated power flows across national borders through the world's media and informational channels. And as it flows it is both performed and can be undermined (Thompson 2000: 246–8). An incumbent's 'good name' (Clinton), the 'brand' of a state (US and its refusal to sign up to Kyoto/Bonn) or the 'brand' of a corporation (Nike), all constitute extremely powerful *and* yet extremely vulnerable symbolic capital. As power is increasingly exercised through the performativity of character, brand and name, these are threatened and destroyed through exposure and shame. As scandals get instantaneously transmitted across the globe they threaten such power. Those incumbents, companies or states, which live by such power, can also, through mobile, irreversible, fluid and far-from-equilibrium processes, die a lingering, transparent and utterly captivating death from it. Attempts at producing order spin out-of-control and destroy most of those caught up in the swirling vortex of scandal that can drag down the powerful and much else besides.

Balkin moreover examines how the occurrence of specific scandals is generating a more general 'culture of scandal' within highly mediated capitalist societies, especially the US. He describes this irreversibly developing culture as being like a 'mutating virus', constantly changing its features in order to grow more widely and to spread more quickly (Balkin 1999: 406). Scandals produce new kinds of TV programme, new modes of journalistic reporting, new Web sites, new modes of media competition. These have the unintended emergent effect through positive feedback that: 'Like a particularly obnoxious weed in a field of grass, the culture of scandal gradually pushes aside other discourses and threatens to consume a greater and greater share of public attention, public discussion and public opinion' (Balkin 1999: 207). So what began as the democratic attempt to reveal the many transgressions of the powerful unpredictably turns out to produce a culture that can drive out almost all forms of media reporting that do not display and enhance the virulent culture of scandal.

## Conclusion

This discussion of scandal shows that informational and mediated power is mobile, performed and unbounded. This is its strength and its vulnerability. Attempted ordering even by the most powerful can result in an array of complex unintended effects that take the system away from equilibrium. In such unpredictable and irreversible transformations, power and especially mediated power, is like sand. It may stay resolutely in place forming itself into clear and bounded shapes, with a distinct spatial topology, or it may turn into an avalanche and race away sweeping much else in its wake.

Complexity thus brings out how what Bauman terms 'liquid modernity' is both unpredictable and irreversible, full of unexpected time-space movements away from points of equilibrium. Those caught up in contemporary global scandals are well aware that they are on a nightmare journey through space, a novel space odyssey of the new world disorder.

## Note

1   This is a lightly edited version of the paper given at the Space Odysseys Conference held in Roskilde in late 2001. I am very grateful for the comments of various friends made at that event held shortly after September 11, 2001. Many of the arguments here are explored in much greater detail in Urry 2003.

## References

Arthur, B. (1994) 'Summary Remarks', in G. Cowan, D. Pines, D. Meltzer (eds) *Complexity, Metaphors, Models and Reality*. Santa Fe Institute: Studies in the Sciences of Complexity Proceedings, vol. 19.

Bales, K. (1999) *Disposable People. New Slavery in the Global Economy*. Berkeley: University of California Press.

Balkin, J. (1999) 'How mass media simulate political transparency', *Cultural Values*, 3: 393-413.

Bauman, Z. (2000) *Liquid Modernity*. Cambridge: Polity.

Byrne, D. (1998) *Complexity Theory and the Social Sciences*. London: Routledge.

Capra, F. (1996) *The Web of Life*. London: Harper Collins.

Capra, F. (2002) *The Hidden Connections*. London: Harper Collins.

Cilliers, P. (1998) *Complexity and Post-Modernism*. London: Routledge.

Dicken, P., Kelly, P., Old, K., Yeung, H. (2001) 'Chains and networks, territories and scales: towards a relational framework for analysing the global economy', *Global Networks*, 1: 89-112.

Dillon, M. (2000) 'Poststructuralism, Complexity and Poetics', *Theory, Culture and Society*, 17: 1-26.

Foucault, M. (1977) *Discipline and Punish*. London: Allen Lane.

Gitlin, T. (1980) *The Whole World is Watching*. Berkeley: University of California Press.

Gray, J. (2001) 'The era of globalisation is over', *New Statesman*, 24th September.

Gunaratna. R. (2002) *Inside Al-Qaeda. Global Networks of Terror*. New York: Columbia University Press.

Habermas, J. (2001) *The Postnational Constellation*. Cambridge: Polity.

Klein, N. (2000) *No Logo*. London: Flamingo.

Knorr-Cetina, K. (1997) 'Sociality with objects: social relations in postsocial knowledge societies'. *Theory, Culture and Society*: 14: 1-30.

Latour, B. (1993) *We Have Never Been Modern*, Hemel Hempstead: Harvester Wheatsheaf.

Latour, B. (1999) 'On recalling ANT', in J. Law and J. Hassard (eds) *Actor Network Theory and After*. Oxford: Blackwell/Sociological Review.

Law, J. (1994) *Organizing Modernity*. Oxford: Basil Blackwell.

Lukes, S. (1973) *Power. A Radical View*. London: Macmillan.

Lyon, D. (2001) 'Surveillance after September 11', *Sociological Research Online*, 6 (3).

Mahoney, J. (2000) 'Path dependence in historical sociology', *Theory and Society*, 29: 507-48

Meyrowitz, J. (1985) *No Sense of Place*. New York: Oxford University Press.

Murdoch, J. (1995) 'Actor-networks and the evolution of economic forms: combining description and explanation in theories of regulation, flexible specialisation, and networks', *Environment and Planning A*, 27: 731-57.

Plant, S. (1997) *Zeros and Ones*, London: Fourth Estate.

Power, M. (1994) *The Audit Explosion*. London: Demos.

Prigogine, I. (1997) *The End of Certainty*. New York: The Free Press.

Ritzer, G. (1998) *The McDonaldization Thesis*. London: Sage.

Sayer, A. (2000) 'System, life worlds and gender: associational versus counterfactual thinking', *Sociology*, 34: 705–25.

Swyngedouw, E. (1992) 'Territorial organization and the space/technology nexus', *Transactions, Institute of British Geographers*, 17: 417–33.

Thompson, J. (2000) *Political Scandal. Power and Visibility in the Media Age*. Cambridge: Polity.

Urry, J. (2000) *Sociology Beyond Societies*. London: Routledge.

Urry, J. (2003) *Global Complexity*. Cambridge: Polity.

Urry, J. (2004) 'Small worlds and the new "social physics"', *Global Networks*, 4: 109–30.

Waldrop, M. (1994) *Complexity*. London: Penguin.

Weiss, L. (1998) *The Myth of the Powerless State*. Cambridge: Polity.

# Chapter 2

# Ambivalent Spaces of Restlessness: Ordering (Im)mobilities at Airports

Jörg Bechmann

## Introduction

Why are there rules and regulations for mobility? Why are there traffic signals and sign-posts that tell us when to stop and when to go? Why are there roads, cycle-paths and pavements each reserved for one particular type of mobility? The answer to these questions, most likely given by both transport professionals and 'lay' drivers alike, might be something like this: 'Because otherwise we would have to cope with unbearable chaos. We would have people and things all over the place – dislocated, unidentified, roaming travellers that criss-cross each others' paths at free will. In the case of traffic, we would have chaotic urban conditions, exorbitant death tolls, major economic crises and numerous other problems. We would approach the end of the "civil society of automobility" (Sheller/Urry 2000). Everyday mobility would no longer be possible'.

In all probability these visionaries are perfectly correct. But then – what about Italy? What about a traffic system that appears to Germans, Swedes or Britains as a 'responsible anarchy', as messy, but nevertheless somehow functional? Is it that different mobilities can indeed coexist without one immediately subordinating the other? Is it that once we abandon traffic rules and regulations, people will start negotiating their right of way and somehow develop informal rules and tacit knowledge which brings them around just as quickly and safely? Is it that the answer anticipated above was too rushed and that there is more to ordering mobility? Well, then, let's push this first answer aside for a while, and try to look out for some others. Answers that may tell us more about why modern cultures invest such efforts into structuring and ordering mobilities and why they are so seemingly obsessed with marking roads, drawing maps, devising hypertext transmission protocols and stamping passports. So, why are there rules and regulations for all sorts of mobilities?

In order to answer this question, I will pursue two points. The first part of this chapter turns to a well-established sociological category – ambivalence. I consider the notion of ambivalence as crucial for a better understanding of why modern mobilities are dealt with the way they are. Thus, I go on to briefly exploit some of the key-texts on ambivalence and relate their author's arguments to 'mobility'. Here, I will describe the various ways in which (sociological) ambivalence contributes to a better understanding of mobility.

In the second part of the chapter I will follow some of the proposals that were developed in the previous part and turn to a particularly suitable site for the study of ambivalence and mobility – the airport. At any given airport, clusters of movement and non-movement stretch over the vast networks of passenger travel, immigration, tourism, goods transport, homelessness, drug-trade, safety and security as well as many other kinds of mobilities and immobilities. I argue that by wandering the architectural, organisational and legal 'decks' of an airport, one will be able to see clearly how ambivalence shapes some central aspects of the social reality of travellers. It is particularly at the airport, that we are able to grasp hold of a number of threads which weave that seamless web of ambivalence and mobility. In order to show how this web is 'constructed' and 'administrated', I take a closer look at some of its central nodes. In other words, I investigate how mobility and ambivalence are dealt with in the departure, transit and arrival lounges of Copenhagen airport. Here, I am concerned with the tactics of airport managers to structure the travellers' mobilities.

## Ambivalence

Who has not experienced that awkward feeling which surfaces the moment one enters an aircraft. The unease that troubles the traveller when passing through the carrier's door: 'Should I really do this, should I really invest trust into this awesome technology and ignore what I read about its vulnerability, should I really put my fate in the hands of the pilot and forget the stories I heard about the "naked pilot" (Beaty 1995) – now, *should I stay or should I go?*'.

What characterises moments like this is a tension between at least two opposing forces. While boarding, the passenger's mind oscillates between two possibilities – the need to go and the wish to stay. Before being airborne we often find ourselves in a state of ambiguity. We are psychologically ambivalent about the anticipated trip. The question is whether *to fly or not to fly.*

Psychological ambivalence is a state of being that characterises many of the moments that make up the (life-)journeys of modern individuals. 'Should I stay or should I go' is the core question for those whose whereabouts are constantly in transition. The tourist, the vagabond, the migrant, the refugee, the global nomad, they all face the same question for different reasons. But whatever their answer may be, it will not be permanent. Once, a destination is reached, a goal achieved, the question is posed anew. Time and again they have to make up their ambiguous minds on whether to stay or go. 'Travel', 'movement' and 'mobility', the above examples insinuate, may all be explored in terms of the ambiguities they rely on and cause. Ambivalence, I suggest, may be one of the key concepts to unravel the social nature of mobility and may provide for a partial answer to the enduring cardinal question of mobility sociologists, that is, why do people move. In the light of this suggestion, it is my intention to show that mobility is about ambivalence – and vice versa.

The introductory example provides the point of departure for weaving together mobility and ambivalence. There, we encounter a psychologically ambivalent airline-passenger. Despite the psychology involved, what is at work here can also be scrutinised with respect to social structure. In that case, it is sociological ambivalence that causes social action to oscillate between two poles. Whereas psychological ambivalence refers to the personality, 'sociological ambivalence refers to incompatible normative expectations of attitudes, beliefs and behaviour assigned to a status (i.e. a social position) or to a set of statuses in a society' (Merton/Barber 1976: 6). Just like inconsistent feelings and desires, conflicting social norms leave social actors in a state of ambivalence.

Now, one way to tie mobility to ambivalence would be by viewing the former as a strategy or practice to minimise the latter. Mobility, for many, seems to perform the function of disabling ambivalence and enabling univalence. It allows the (temporal) escape from a place where individual existence is torn between opposing needs and norms. In his classical example of how the structure of social relations may foster ambivalence, Robert Merton together with Elinor Barber has implicitly touched upon these enabling capacities of mobility. They write: 'For example, the relation between master and apprentice – say in the world of science – may be one in which, for structural reasons such as a paucity of major chairs in the field, the apprentice "has no place to go" after having completed his basic training, other than the place still occupied by the master. This is one type of structural situation; the one in which ambivalence is apt to develop. But if the structure of the society provides an abundance of other places, some as highly esteemed as that currently occupied by the master, the apprentice may be less motivated for these structural reasons to develop ambivalence towards the master' (Merton/Barber 1976: 5).

Having 'no place to go' produces ambivalence, 'an abundance of other places', though, reduces ambivalence, that is what Merton and Barber are saying here. In a global society where more and more people are said to have more and more places to go, global ambivalence, one might assume, would shrink and universal univalence would rise to an all-time high. If univalence is understood as the existence of a unified and universalised set of norms and values, then this assumption matches pleasantly with a multiplicity of modern utopias of 'The One World', where travel allows for 'diversity' to be replaced by 'unity'. More mobility, so a popular version of the virtues of travel, diminishes cultural boundaries and unifies cultural diversities. Here mobility is the terminator of ambivalence.

The equation 'more mobility equals less ambivalence', though, seems to contradict common knowledge. What common people know is that mobility-gains often materialise as ambivalence-growth. Being 'highly automobile' and owning more than one car makes choosing which one to drive both necessary and difficult, being residentially mobile and having lived in lots of places, makes it difficult to decide where to stay, being sexually mobile and having been promiscuous makes being monogamous sometimes a problem. The psychological ambivalence of all of these mobile individuals certainly has structural roots. Not really being able to decide whether to stay or go, I claim, is a consequence of the social structures that underpin any kind of mobility.

What is it that constitutes these structures of mobility that evoke ambivalence? One important structural feature emerges from the dual meanings of a journey's origin and destination. Every trip has two poles, both of which equally attract and reject the traveller. Against a mechanic modelling of movements, which suggests a journey's origin to be 'pushing' and the destination to be 'pulling' the traveller, I claim that both origin and destination equally push and pull. Tourists, for instance, are just as much irritated by the 'cultural differences' of their destination as they are fascinated by them. Refugees are just as much looking forward to escape devastation and oppression, as they are wary to leave the community of friends and family. For the array of post-modern nomads who criss-cross the globe, ambivalence is a reoccurring state of being. Ambivalence is already inscribed into their social roles through the ambiguous meanings of two places. With this, the former suggestion that mobility terminates ambivalence can no longer be upheld. Rather I would suggest that increasing mobilities and growing ambivalences are viewed as twin processes – the more we travel, the more ambivalence there is.

But let us take this a bit further and place both mobility and ambivalence into the modern condition – let us weave together modernity, mobility and ambivalence. Amongst contemporary social theorists, modernity is generally held to be a restless epoch. Modernisation unleashes and accelerates various mobilities. The continuous movement of capital, people and information, for whatever reason, dismantle geographical, political, social and cultural borders and produce all-pervasive 'spaces of flow' (Castells 1996). This tradition of viewing modernisation as tantamount to mobilisation spans from the authors of the communist manifesto for whom 'all that is solid melts into air' to Zygmunt Bauman's 'liquid modernity' (2000). At first sight any of the prominent modernisation theories seems to be a mobilisation theory in disguise. Modernisation is tantamount to mobilisation. The more modern we are, the more we travel.

But what about ambivalence and modernity? Three authors, Donald Levine, Niklas Luhmann and Zygmunt Bauman, might offer some initial help to answer this question. For Levine (1986), to begin with, modern societies, unlike traditional societies practice a continuous flight from ambiguity. He claims that the 'institutions and ideals of modern culture are seriously dependent on unambiguous modes of expression. Modern science, commerce, occupational specialization, bureaucratic management, and the formal rationalization of legal procedures and much else are all unthinkable without resources for clear and distinct communication' (Levine 1986: 38). Here, Levine turns to language in order to unravel how modernisation is entangled with manifold attempts to reduce ambivalence.

For Luhmann, modern societies rest on functional subsystems. 'Funktionssysteme', such as politics, science, economics, medicine etc., structure themselves not only by their very own interpretations of what is right and what is wrong, but they also do this through a rigid application of binary codes. 'Die Besonderheit funktionaler Teilsysteme ist, dass sie ihr Beobachtungsschema über die strikte Zweiwertigkeit ihrer binären Codes generieren. So ist für Politik entscheidend, ob man Amt und Entscheidungsmacht innehat oder nicht, für Wirtschaft, ob man zahlt oder nicht, für Religion, ob etwas dem Heil oder dem moralischen Standard dient oder nicht, für Erziehung, ob etwas im Hinblick auf Chancen im Lebenslauf gelernt wird oder nicht'[1] (Knerr/Nassehi 1993: 132). Here, late modernity is described as a societal formation that creates order through demanding a choice between two opposing options. Ambivalence, that is to say the oscillation between these binary poles, is not an option.[2]

For Bauman, just like Levine and Luhmann, modernisation seeks to end ambivalence. He writes: 'The typical modern practice of the state, the substance of modern politics, was the effort to exterminate ambivalence: to define precisely – and to suppress or eliminate everything that could not or would not be precisely defined' (Bauman 1990: 165). This, though, is only half of Bauman's story. The other half begins where the first one ends. By introducing the reader to a 'stranger', who represents the anomalous category between our 'friends' and our 'enemies', Bauman shows how modernity's 'horror of indetermination' turns against itself. In other words, the eradication of ambivalence under the modern order, re-solidifies what it seeks to dissolve. Ever more strangers, as the epitome of any unwelcome 'undecidables', have come to populate the modern state as attempts were made to absorb them under the friends/enemy dualism. Bauman maintains that 'the strangers refused to split neatly into "us" and "them", friends and foes. Stubbornly, they remained hauntingly indeterminate. Their number and nuisance power seem to grow with the intensity of dichotomizing efforts. As if the strangers were an "industrial waste" growing in bulk with every increase in the production of friends and foes; a phenomenon brought into being by the very assimilatory pressure meant to destroy it' (Bauman 1990: 155)

What does not work in relation to the unambiguous attempts to assimilate strangers, does not work for any modern flight from ambiguity. No matter what kind of flight, trip, journey, displacement we are looking at, *ambivalence is both its cause and consequence*. Take the above apprentice: The mobility of Merton's and Barber's apprentice may have temporarily freed him from some of the more problematic personal consequences of his ambivalence towards the master. However, by embarking on that journey, his ambivalence did not vanish. Rather, by leaving the master's territory, the apprentice offered his ambivalence some reproduction-space. By moving away, the apprentice assures himself the possibility of living (with) his ambivalence. By having 'a place to go to' he does not alter the structures upon which ambivalence unfolds. The ambivalence-causing principles of master- and scholarship are revitalised, when, in another place, the apprentice becomes a master himself. Now, he himself will meet the flip-side of his former ambivalence towards his master – in a double sense. With upward social mobility the ex-apprentice will be the projection for another apprentice's ambivalence, and he will respond to that by being equally ambivalent towards his new apprentice. When the apprentice flew from ambiguity he made sure that it continued to exist.

As we have seen through the eyes of Zygmunt Bauman, modernisation is about ordering that which does not fit the order. In fact, it is about ordering that which cannot be ordered. But in order to create a univalent 'order of things', all 'things' have to be moved – movement of people and things for the sake of creating order and preventing disorder. But from the point of view of the ordering institution this very movement is of a mixed blessing as it reanimates what it wished to destroy.

Against this background, it now seems unintelligible to argue that as modernity progresses, mobility increases and ambivalence declines. Instead the equation would run like this: *as modernity progresses, mobility increases and so does ambivalence*. Here, mobility is still the strategy to cope with ambivalence. We move away, when we have lost control over living with contradictions. But in doing so we reproduce them. With the cyclic uprooting and re-rooting of modern people, with their repetitive 'coming' and 'going', with their rhythmic engagement and disengagement, their ambivalence is flourishing. With every flight from ambiguity, ambivalence is growing rampantly.

Clearly, behind the notion of ambivalence some clues that contribute to answering the questions of *why mobility is ordered* and *why there are rules and regulations for mobility* are hiding. These clues suggest that the objective of ordering and regulating mobility is not so much to diminish immobility, but to abandon ambivalence, that is to prevent an ambivalent stage in between mobility and immobility. Traffic lights at urban intersections may serve as the arch-examples to illustrate this very exclusion of ambivalence. Two of their genuine features are important here. First, they are either red or green (and yellow only in order to prepare the driver for a forthcoming phase of 'red' or 'green'). Second, they mobilise pedestrians in the same way as they immobilise cars. On this view the traffic light signifies modernity's inherent drive to eliminate 'undecidables' and to construct order through dualisms.

But this attempt to abandon ambivalence is somewhat of a hopeless enterprise in a society that is characterised by all-pervasive mobility. Under the condition of mobility, ambivalence cannot be abandoned, because with any new 'movement'/'non-movement' more ambivalence and uncertainty is generated (remember Merton's and Barber's apprentice). Again, this impossible flight from ambiguity is most apparent in traffic. It seems that any attempt to order everyday mobility creates new disorders, every transport solution produces another transport problem. It is therefore, that transport solutions proposed and implemented by yesterday's transport experts have transformed into contemporary transport risks. Today's risks can be seen as the consequences of the problem-solving efforts and the 'quest for order' of modern transportation planners. Such quests for order, inherent in the scientific risk rationalities of many transport researchers lead to increasing ambivalence, uncertainty and contingency – in short, more 'problems'. As such, 'problems are created by problem-solving, new areas of chaos are generated by ordering activity. Progress consists first and foremost in the obsolescence of yesterday's solutions' (Bauman 1991: 14). In the case of automobility, there was the task of improving transportation which was resolved thanks to the automobile. Then there was the task of increasing road capacity in urban areas – resolved thanks to the development of inner-city highways and ring roads. Then there was the task of improving drivers' safety – resolved thanks to a more sophisticated road design and the safety belt. Then there was the task of protecting the pedestrian from the driver (whose ensured protection made him/her more dangerous to other transport users) and so on and so forth. It seems as if each new transport solution creates a number of consecutive problems.

However, all this does not reveal much about the implementation of an order that seeks to eliminate ambivalence. It says nothing about *how mobility is ordered*. An answer to this question is not derivable from mere theoretical reflections. The clues to unravel how mobility is ridded of ambivalence can only be found in the *field*, that is to say through an empirical analysis of *situated mobilities and order-making processes*. I would like to outline some initial steps of such analysis now. Therefore, in the second part of this chapter, I will consider an international airport as a site at which I believe some of the hidden links between contemporary mobilities and ambivalences can be uncovered.

## Airports

The above section illustrates two things. Firstly, it reveals mobility's proximity to ambiguity. The argument, which I am advancing here, is that the growth of virtual, corporeal, residential, religious, sexual etc. mobilities as well as their ever-tighter coupling produce ambivalence. Secondly, as these complex mobilities continue to foster ambiguities they become the core concern of all kinds of mobility managers. A number of modern professions, organisations and industries are guarding and patrolling such mobilities. They seek a stricter mobility-order in so as to hinder the production of ever more ambiguities, uncertainties and contingencies. The crucial question now is how do they do this? What are the ordering practices of those who fear that an interweaving of too many mobilities create too many problems? Possible answers to these question can, for example, be found on the desks of immigration officers, telecommunication engineers or traffic planners. Studying their specific positions and multiple relations within the actor networks of immigration, telecommunication or traffic would provide an insight into the practices of ordering modern mobilities.

Another possible way of arriving at an answer to the above questions would be to explore the artefacts of mobility. The most popular example for that sort of exploration is the automobile. The car has been turned inside out and upside down multiple times by scholars of (auto)mobility. They have shown how automobility has become ambiguous, produced unintended consequences and, then, turned against itself (Urry 2000, Rammler 2001, Kesselring 2001, Beckmann 2001). In the second part of this section, I will follow in their footsteps, but instead of travelling with, what Paul Virilio would call, a 'dynamic vehicle', I will visit a 'static vehicle', that is to say an international airport. By addressing airports as sights for social studies of mobility, I seek to better understand how the guardians and patrollers take ambivalence out of mobility.[3]

The vastest of such airports are now considered hubs. They are the focal points of a global air-traffic system that spans from remote regional runways like Billund International Airport to mega-structures like that of LAX, CDG or LHR. But more than that they are also the kind of hubs at which an array of diverse mobilities intersect.[4] The co-presence of cargo, tourists and refugees as well as sounds, pollution and images on the move are posing colossal challenges to the skills of all sorts of mobility managers. At the airport, these mobilities are exposed to rigid de- or accelerations depending on whether they are starting or landing, unwanted or welcome, secure or unsafe. Clearly, the legal, technical and organisational fabric of international airports are permanently handling, steering, channelling, protecting, prohibiting and managing multiple mobilities rather than just getting people and goods 'in the air' and 'on the ground' again. It is this multiplicity of (im)mobilities that renders ambiguity the central theme, which is inscribed in and signified by the very architecture of any international airport, as Pascoe says: 'Architecture, obviously, is an art of immobility, of frozen time, of suspended movement, and in a society where transit looks like flight, and where perpetual motion abolishes places, its practitioners seek to affirm the values of stability, identity, presence, by resisting movement, by preventing everything becoming an indistinct flow. Such an affirmation is not a negative reaction; rather, it is mediation between flight and confinement. Existing with the aegis of such an antithetical concept, airports should never be taken for simple thoughts; they are neither monuments to immobility, nor instruments of the mobile society, but instead the improbable conjunction of both' (Pascoe 2001: 14).

What Pascoe shows us here, is that the multiple functions and actions performed by and at airports render the terminal an ambivalent space. At the terminal, ambivalence is not only a state of mind, a feeling that surfaces, while we cue for boarding. Rather, it is a central brick in the structure of any 'airspace'. The airport forms a(n) (infra)structure called ambivalence. Its in-built ambivalence paves the ground for human behaviour to oscillate between staying or going.

By standing in and reading Pascoe's *Airspace*, we are able to rediscover certain aspects of the texts by Merton and Barber, Levine and Bauman. With reference to these theorists, I suggest that the structure of the airport represents the sort of ontological ambivalence one may presume for any kind of social world. Due to the coexistence of the diverse 'demands' forwarded by the ambivalent airport-structure, the traveller is coerced into oscillating between multiple roles. The ambivalence of the passenger is, thus, not only a question of the structure of travel, where both origins and destinations are equally pushing and pulling, but equally rooted in a transport infrastructure like the airport.

Now, the question that evolves from this is how are certain 'air-mobilities' dealt with, organised and managed? What are the strategies of airport operators to order the diverse and ambivalent mobilities? To answer these questions, I shall first return to the ambivalent passenger in the introduction above. Hence, the basic idea that frames this makeshift inquiry runs as follows: Picture the above airline-passenger arriving by train at the airport, in this case Copenhagen Airport. Between leaving the train and becoming airborne the anticipated passenger will pass through a number of 'thresholds' – the parking-lot/bus- or train-station, the terminal entrance hall, the check-in, the security-check, the boarding-pass control, the aircraft door, to name just a few. These distinct points mark the passenger's journey from airport-entry by land to airport-exit by air. As s/he moves from one threshold to the next, the passenger is guided through the transit area by several layers of signs invented and erected for the purpose of ordering diverse and often conflicting mobilities. Let us take a closer look at this journey and explore how it relates to the notion of ambivalence.

*The arriving and departing passenger*

The arrival at Copenhagen Airport (CPH) by train immediately situates the traveller into a 'messy' space. The area above the platforms hosts both check-in counters for the departing passenger and a congregation area for the departing passengers and those who have been waiting for them. Originally, this part of the airport was meant to house just a few 'check-ins' for departing passengers arriving by train to CPH. Now it is a full-blown terminal, in which arrival traffic and departure traffic are intersecting. But despite the mess of movements there, this intersection is an ordered one: from the door that spills out those arriving the floor is marked with a black alley – suggesting to the arriving passengers that this is the path to follow. None of the spouses, friends and colleagues waiting, dare to stand in this blackened area. As those waiting are separated from those arriving, the departing passengers are guided away from this space of ambivalence. Additionally, there are signs and monitors that address both of these groups and render their mobilities into streamlined movements forward.

Here one already encounters the essentials of indoors-mobility management at airports – architecture and audio-visual sign systems. Together the 'signs and spaces' of airport terminals structure the passenger's movement forward. They are somewhat universal and take on similar features in most of the world's major airports. To be found in any airport, these main-frames for movement, make getting lost most unlikely.

*The passenger in question*

At the next threshold, the check-in counter, the passenger is questioned. A few questions by the ground-personnel will bring to the surface whether he or she is ambivalent about the forthcoming trip. Skilled flight attendants and experienced check-in personnel are able to filter out 'ambivalence'. The filtering mechanism is based on detecting mismatching features – a feature that does not really match the profiles. A drop of sweat, nervous laughter, awkward gestures …

This ability of detecting the ambivalent passenger is not so much a question of training, but rather one of professional experience and personal judgement. A couple of 'normal' questions ('Do you have any luggage to check-in?' 'Where would you like to sit?' 'Are you aware of our frequent flyer programme?' 'Would you like to join it?') may create clarity about what troubles the passenger in question. But questioning the passenger has more purposes than knowing what is wrong with him. For the anxious traveller being questioned is tantamount to being taken care of. There is a feeling of safety evoked by showing interest into the passenger's wellbeing – 'we ask a question, because we care for you'. Ambivalence vanishes the moment I see that someone takes my worries seriously and begins to ask questions. Every single one of those questions, of course, calls for an answer. In order to reply to the question I have to make up my mind on whatever I wish to answer. The more answers are given by the passenger, the more his ambivalence may shrink. Answering a question, by taking a decision demands engagement and acceptance. Once engaged in organising the check-in, boarding, etc., there may be less room for ambivalence to emerge.

Clearly, at some thresholds the passenger will have to respond to questions. At others s/he will have to make distinct choices between different options. For his or her own sake the responses and decisions asked from the passenger must be univalent. To respond to questions like the ones above by uttering 'I don't know' (… whether I have packed the suitcase myself), will certainly disrupt the journey from platform to terminal. Clearly, the trip through the airport reminds one of an examination. What is being tested here is the capability of the passenger to decide. What is at stake here is the termination of the traveller's ambivalence as s/he wanders through the thresholds.

*The filed passenger*

If a case of ambivalence is detected an electronic file is created at the check-in. Notes are added to this virtual travelogue as the passenger progresses through the airport. Every time the traveller contacts airline personnel at the information counter or boarding desk another note enters the record. The airline creates somewhat of a virtual passenger that takes account of the real passenger's ambivalent features. Normally, the file is deleted as the plane takes off. The virtual ambivalent passenger neither leaves the ground, nor the check-in-system. Only in cases of severe ambivalence will the file be transferred to the reservation system. For a couple of hours, the virtual passenger then becomes the real passenger's travel companion. This airline practice is not so much concerned with immediately diminishing ambivalence, but rather about preparing for it. Knowing the history of ambivalent behaviour may ease later attempts to tackle it.

### *The stripped passenger*

At the check-in counter another performance takes place. Here, the 'trip through the airport' starts to turn into a 'strip at the airport'. Stripping and tripping are the twin activities that the passenger performs as he or she moves to the gate. Giving away the luggage, emptying the pockets, being deprived of the ticket and left with the boarding pass and finally having to place the handbags in the overhead-locker during take-off and landing – these are all incidences, where the passenger is being stripped of something. With every hand-over the traveller passes on a piece of him/herself. Once in the hands of the airport and airline personnel, the traveller will sense and sometimes experience the increasing barriers for possible outbreaks of ambivalence. By being shorn of much of his or her belongings the traveller is increasingly confined in his or her options to express ambivalence. S/he has given vital parts to the airport and airline, which in case of an outbreak of ambivalence, will be difficult to reclaim – the money spent on the ticket will most likely be lost, the reputation gone and future access made more difficult.

What the giving away, handing over, and passing on of personal belongings signifies, is an act of trusting. The rites of passage that take place at every threshold are meant to make the person less ambivalent towards the forthcoming flight. Their intention is to strengthen the bond between the individual traveller and the air-traffic system. Every passage tightens the mutual dependency of agent and system and eases the establishment of trust. From the airport-LTS comes the message 'if you allow me to take care of your belongings you will have little to worry about and be safe'. With the passing on of things, trust comes and ambivalence goes.

### *The cordoned passenger*

Having passed a number of thresholds in the departure area, the passenger enters the 'transit'. Here, his mobility is cordoned by the 'main flow'. The 'main flow' is running along a 6 meters wide alley, which spans over major parts of the transit area. The alley itself is reserved for guidance systems exclusively. Its sides are housed and cordoned by a number of shops.[5]

The key task at the main flow is to order both univalent and ambivalent movements. There is an ideal univalent movement from luggage control through the transit area to the gate, and there are ambivalent movements from shop to shop criss-crossing the main-flow. Both movements call for their very own guiding systems, which seem to interfere frequently. (This situation becomes even more complex as an array of new 'vehicles' – such as scooters of all sorts – enter the pedestrian's main flow.)

The conflicting interests between the 'flow managers' and the 'shop managers' structure the movements of passengers. As the first call for more univalent movement through the transit area, the latter are in favour of more ambivalent motion. The overall result is a heightened ambivalence in an area, which was originally created as a cordon of signs leading the passenger straight to the gate. This heightened ambivalence is illustrated by the various advertisements invading the 6 meters wide path to univalence.

*The delayed passenger*

Delays cause problems, because the journey has neither ended, nor can it continue. These ambivalent moments create costs for both airlines and airports and they leave the airline passenger dissatisfied, frustrated and angry. Hence, efforts are made to reduce delays or manage them better. One way of dealing with interruptions in the flow of passengers is to alter the perception of waiting time by, what could be called, 'time-lapsing techniques'. Engaging the waiting passenger in shopping, eating, reading, talking, playing etc. are ways to bridge those moments in which mobility is disrupted. Delay-management techniques like these are often applied in situations where movers are temporarily denied their movement. 'Queuing-entertainment' in front of theme parks or the 'Stauberater' (Engl.: traffic-jam consultant) on the German Autobahn are such techniques.

However, as contemporary delay-management techniques aim at lapsing waiting-time, they actually cause another kind of displacement of the passenger. Infotainment screens and www-access in terminals allow for imaginative and virtual travel. Silent areas and personal corners equipped with audio-visual systems allow for retreat and escape from the crowed spaces of temporal immobility. Spatio-corporeal immobility is in this case battled with mobilities of the mind. What the flow managers at airports offer are exchangeable mobilities in case of delays. Real mobility is substituted by virtual mobility, so that the passenger somehow keeps on moving. Moreover, that sort of surrogate mobility is of a different character than the actual linear movement of the passenger from check-in desk to aircraft-seat. The stand-in mobility performed by the traveller is like a drift without getting lost. The passenger-as-drifter is allowed a controlled escape. Drifting through the virtual and imaginary cyber- and dream-scapes the delayed passenger is in a stand-by position until further notice given by audio-visual information-systems.

*The assimilated passenger*

If ambivalence erupts once the passenger has already embarked on the aircraft, the airline begins to assimilate the passenger with other '*nods*' in the actor network of air-travel. 'Seeing the cockpit' and 'meeting the captain' are common ways to assimilate.[6] Moreover, assimilation, in some cases, may need removal. This can be either constant or permanent removal from the aircraft, or re-seating the passenger (though never to a seat in between two others). The latter form of removal, again, serves the need for assimilation. The new seat will bring the passenger closer to the flight-attendants. From his new seat he can both be seen by the crew and see how competently the crew performs its multiple tasks.

To summarise the main findings, the passenger's journey from the train-platform through the air-terminal to the boarding-gate at CPH is meant to be a unidirectional one. It is designed as a one-way trip. However, this journey takes places on a platform, which allows as well for more ambivalent movements. Apart from moving univalently and directly from check-in to gate, the passenger enters spaces of ambivalent mobilities. In order to assure that univalent movement endures even under the condition of ambivalence, the flow managers at CPH have erected signs, invented questions, created check-points etc. They have established a system of mobility where every threshold presents another 'point of no return'. The further the passenger advances and the closer s/he comes to her aisle- or window-seat, the harder will it be to reconsider and *decide to stay*. The passenger's approach through the airport to the plane is like a trip down a narrowing pipe, leaving less and less space to stop and turn. Once certain thresholds are passed, the ability to express ambivalence (and maybe turn around) is diminishing. Once the traveller has checked-in, purchased a bottle of duty-free Scotch in the transit area and started boarding, a sudden eruption of ambivalence will create subtle disturbance amongst both airline personnel and fellow travellers.

## Conclusion

In this chapter I have sought to do two things. Firstly, I have shown that mobility produces ambivalence and that modern societies are preoccupied with eliminating ambivalence. With respect to the multiple coexisting mobilities that constitute modernity, this elimination of ambivalence results in rigid ordering-programmes. By ways of ordering movement and motion, actors seek to take ambivalence out of mobility. Secondly, I have explored a particular site, where such ordering is most prominent – the airport. To order means to guide, question, file, strip, assimilate and remobilise the passenger.

Here, mobility is deprived of ambivalence by ways of *separation*. What happens at the airport is that different mobilities are separated from each other. This is reflected by both the architecture of the airport and its organisation: there are cues for different air-lines and destinations; the mobility of air-travellers is separated from that of their luggage; on-board virtual mobility by mobile phone is overridden by spatial mobility; during take-off and landing walking up and down the aisle is not possible.

Separation, one could argue, is the overarching purpose of any modern mobility management.[7] In order for different mobilities to co-exist, they have to be separated. Or, to employ a systemtheoretical phrase, mobilities are to be differentiated into various sub-mobilities. On this view, it is Luhmann's systemtheory that casts some light on why and how mobilities are ordered through separation. A strict binary coding – mobility *vs.* immobility, or, mobilisation *vs.* immobilisation – is the arch-instrument for the flow and mobility managers at any international airports

Against this background, mobility managers are now endowed with a twofold task: On the one hand they need to separate and disentangle these complex interwoven mobilities: luggage from people, cars from open spaces (parking-lots and runways) and virtual from spatial mobility. On the other hand, they need to reassure that all sub-mobilities retain their ties and are synchronised so that mobilisation can reproduce itself. The managers are in charge of the 'programmes', which decide on the openness of mobility and its sub-systems. Such programmes decide which object or subject has to stop in order for the other to continue; which mobility is slowed down, while the other is accelerated. In the end though, it is this ambivalent *coexistence of mobility and immobility* that characterises the airport. Rather than disentangling mobility and immobility any attempt to do so reinforces a mutual dependency of those binary codes and, thus, de-differentiates sub-mobilities just as much as it differentiates them.

To capture the professional (re)making and juggling of coexistent mobilities and immobilities I would like to end this chapter by advancing another term, that is to say the concept of *motility*. In a first approximation, the notion of motility captures the various stages, in between mobility and immobility. Motility, I claim, means neither immobility nor mobility, neither stability nor instability, neither solidification nor liquefaction. It describes the motile moments when people and things are physically, virtually or residentially not quite at rest and not quite on the move. It portrays the situations in which both objects and subjects are simultaneously in and out of the actor networks that hold them in a fix.

Everyday examples of these ambivalent moments are numerous. One of the most prominent examples is given by the virtual traveller at rest in front of his or her interface. Working at and sitting in front of the home PC and travelling through cyberspace is a mode of being in the world that combines mobility and immobility – here, the cyber-traveller lives a motile existence. An equally motile image is evoked by the ever present 'car-driver hybrid' cruising down the road whilst inert in his seat. Or, motility can be detected in the asylum-seekers camps of the 'Schengen Countries', where the uprooted and dislocated migrant is temporarily fenced in and fixed before being returned 'home' or admitted entrance. For these travellers being in the world is a motile experience.

What these examples illustrate is that despite the emphasis given to mobility in recent social theory, immobilities do hang about. The outcome of such co-presence of mobility and immobility is motility. Hence, exploring the structuring of (im)mobilities, first and foremost, would mean to cast light upon the multiple modes of motility that frame the social embeddings of modern travellers. One of such modes, I suggest, is the construction of safety/security within air-traffic and at airports by ways of ordering (im)mobilities. Or, put in other words, *being safe and secure is tantamount to being mobile and immobile at the same time.*

At any given airport the task of the 'flow managers' is to ensure the immutability of the mobile passengers, suitcase, aircraft etc. under conditions that are highly uncertain. Their jobs are about creating '*immutable mobiles*' (Latour 1988) in a motile world. According to Latour immutable mobiles are travelling entities that although in motion are able to maintain stable relationships and thus remain in effect immutable. This very simultaneity of motion and inertia is illustratively explained by Law and Mol (2001). For them, mutability and mobility make reference to two types of spatialities. Following their interpretation, an airline-passenger would exist in both *network* and *Euclidean* space. To occupy a fixed position in the actor-network of air-traffic means immutability. The traveller would then be safe. Despite its movement through Euclidean airspace, the position in network space is stable. Only as long as this relational positioning can be maintained, people, capital, goods and information can be moved. Or put in other words, mobility relies on the immobility of the agent in network space. On this view, resurrecting security, safety and certainty is about both mobilising and immobilising the subject. It is about introducing *motility* into the agent's ways of being in the world.

## Notes

I thank Anni Dalsgaard from the 'SAS Information Department' and Lars Skov from 'Facilities' at Copenhagen Airport for providing me with helpful information.

1   English translation: 'The peculiarity of functional subsystems is that they generate their obseravation-schemes through strict binary coding. For politics it is decisive whether office and power is one's hands or not, for economics, whether one pays or not, for religion, whether something serves the moral hail or not, for education, whether biographical lessons will be learned or not' (Knerr/Nassehi 1993: 132).

2   If one applies Luhmann transport and mobility systems this binary code would be 'movement'/'standstill'. Traffic, for example, could then became an interesting area for a thorough system-theoretical inquiry which unravels the 'programmes' by means of which the traffic system decides between movement and standstill. To my knowledge such inquiry has not yet been conducted.

3   Below, I shall present the first stage of such a study. The expositions must not be read as results of, for example, an ethnography of airports. Rather, they are products of a 'sociological reflection' enriched with findings from an ad-hoc empirical investigation. As such, they refer to both my own unstructured observations and interviews at airports, and the thoughts, arguments and conclusions forwarded by other 'airport-investigators' (for instance, Gottdiener (2001) or Pascoe (2001)). Consequently, they are of a preliminary character and will be exposed to critical review once this study of the social life of airports is moving ahead.

4   Amongst the recent explorers of airports, it is Pascoe (2001) in particular, who provides some inspiration for a study of ambivalence and mobility at airports. Unlike Gottdiener (2001) who primarily interprets the airport as 'a kind of shopping-mall', Pascoe reads ambivalence into the structure of the airport. Although both speak of airports as cities and conjure up images of urban experiences, Gottdiener's terminal is about *shopping* and *travelling*, whereas Pascoe's airport 'is best viewed not as a discrete structure, but as an array of functionally diverse spaces designed on various scales, a network of ticketing checks, security searches, customs inquiries and transit zones through which food, fuel, mail, cargo, baggage and air traffic are circulated' (Pascoe 2001: 11).

5   Another cordon used to await the arriving passenger. Until recently those arriving at CPH by air were led along an elevated path across the check-in desks in terminal 2. The sides of the cordon were approximately 3 meters high and almost impossible to see through. The point here was to make sure that the arriving passengers would not 'drop off' items before passing the customs after the luggage reclaim. The cordon not only served to guide the arriving passenger straight to the luggage reclaim, but also ensured that the passenger arrives in one piece. Whereas stripping was the thing to do at the departure, gathering and holding onto your belongings is the task at the arrivals.

6   At SAS, this practice has changed after the 11th September. No longer is there be a calming of ambivalent passengers by showing the cockpit and meeting the captain. Moreover, the screening of passenger at the check-in or the gate has intensified. Slight signs of ambivalence may create more intense responses by both the ground personnel and the crew.

7   Enter the www-pages of the Eurocontrol Guild of Air Traffic Services (www.eurocontrol.org) and you will here the jingle 'you got to keep on separating'. Clearly, the overall motto of modern mobility orders.

# References

Bauman Z, 1990, 'Modernity and Ambivalence' *Theory, Culture and Society* **7**, pp 143–169.

Bauman Z, 1991 *Modernity and Ambivalence* (Polity Press, Oxford).

Bauman Z, 2000 *Liquid Modernity* (Polity Press, Cambridge).

Beaty D, 1995 *The Naked Pilot* (Airlife, Shrewsbury).

Bechmann, J, 2001 *Risk Mobility – the filtering of automobility's unintended consequences*, University of Copenhagen, Department of Sociology, PhD dissertation.

Castells M, 1996 *The Rise of the Network Society* (Blackwell, Oxford).

Gottdiener M, 2001 *Life in the air – surviving the new culture of air travel* (Rowman and Littlefield, Oxford).

Kesselring S, 2001 *Mobile Politik* (Sigma, Berlin).

Knerr G and Nassehi A, 1993 *Nikals Luhmanns Theorie sozialer Systeme* (Wilhelm Fink Verlag, München.

Latour B, 1988, 'Opening one eye, while closing the other ... a note on some religious paintings' in *Picturing Power: Visual Depictions and Social Relations* (Eds) G Fyfe, J Law (Routledge, London) pp 15–38.

Law J, Mol A, 'Situating technoscience: an inquiry into spatialities' *Environment and Planning D: Society and Space* **19** (5) 609–621.

Levine D, 1986 *The Flight from Ambiguity* (University of Chicago Press, Chicago).

Merton R K (with Barber E), 1976, 'Sociological Ambivalence' in Merton R K *Sociological Ambivalence and Other Essays* (Macmillan, London).

Nedelmann B, 1996, 'Rober K. Merto's Concept of Sociological Ambivalence: The Florentine Case of the "Man-Ape"' in Mongardini C and Taboni, S (Eds) *Contemporary Sociology* (Transaction Publishers, London).

Pascoe D, 2001 *Airspaces* (Reaktion Books, London).

Rammler S, 2001 *Mobilität in der Moderne* (Sigma, Berlin).

Sheller M and Urry J, 2000, 'The city and the car' *International Journal of Urban and Regional Research* **24** (4) 737–757.

Chapter 3

# Spatiality, Temporality and the Construction of the City

Kirsten Simonsen

Spatiality and mobility are phenomena interfering in the construction of the contemporary world in and across many geographical scales – including the urban one – each opening up different perspectives on their mode of operation. In this chapter, I want to approach the theme from the perspective, not of global networks, but of everyday practices. This choice has a double purpose. On the one hand, I want to transcend, or at least question, some of the most popular images of (post)modern cities in order to develop an understanding of cities and urban culture that recognises the multitudinous character of urban life. On the other, the discussion should be seen as part of an effort to develop an analytical framework from which to approach the cultural construction of the city – or rather 'the multiple faces of the city'.[1] Taking into account these purposes, the chapter is organised in three parts. The first part discusses some prevalent imaginations of the city in popular and academic discourse and suggests rhythmanalysis – as it is expounded by Henri Lefebvre (1992) – as a possible starting point for the development of alternative understandings. The next two sections of the paper proceed from this account, not by suggesting alternative generalisations on the character of urban life, but instead by considering 'decisive' elements in everyday performance and construction of the city/ies.

## Rhythms of the city

A popular account of metropolitan life is one of increasing mobility and speed. Imaginations around that issue is a recurrent motif in both artistic and scientific representations, and more often than not they are based on discourses on a globalised world dominated by a space of flows brought into existence by new information technologies. This space of flows, it is argued, is changing our apprehension of space, time and subjectivity. Places vanish, time becomes instantaneous and the subject becomes decentred, strung out in mobile 'cyborg' identities. A teleological story is told that gives rise to urban nightmares of simultaneity, simulacra and strange 'science-fiction' worlds overwhelming the city and its inhabitants. This emphasis on ceaseless mobility can be found in much contemporary literature, but in particular two authors repeatedly come up in the discussions. The first one is Virilio with his work on speed and the emptiness of the quick (e.g. Virilio 1983, 1991). For his part, we live in 'the age of the accelerator' in which speed is invested with extensive effects. For instance it consumes subjectivity, leading to the disappearance of consciousness and making everyone a passer-by, an alien or a missing person. The other trend-setting work is the one of Deleuze on nomadism; on nomadic subjects that traverses points of possible individuation in a migratory fashion, performing a conductivity that knows no fixed and sedentary boundaries (Deleuze and Guatterie 1988). However, the motif can also be traced back to classical theories of modernity, not least to Simmel's account of metropolitan life in which perpetual changes, overstimulation and the bombardment of increasingly isolated individuals by new impressions, sense experiences and information is a dominant theme (Simmel 1969, orig. 1903)

Notwithstanding the qualities of the analyses conducted within this mobility paradigm, I feel uncomfortable about taking it as a starting point for the analysis of urban life and the cultural construction of the city. I will refer to at least three reasons.

First, these accounts seem to ignore what I would call the 'ordinary', the way in which a lot of urban citizens live an everyday life characterised by human locatedness, routinisation and involvement in different circles of social interaction. The paradigm and the metaphors of mobility, as I see it, in particular becomes a problematic language because it implies not only that we are all on the road, we are also 'on the road "together"' (Wolff 1993). This in itself renders the suggestion of ceaseless and free movements a deception, since we obviously do not have the same access to the road. The objection is not that the dispossessed do not travel but rather that when they do, they do it in different and more constrained ways. In other words, the mobility paradigm needs differentiating socially or, with Doreen Massey (1994), its 'power-geometry' needs considering. This is not a problem concerned with social inequality only, but also with the one of gender. The mobility in question predominantly appears to be a white, middle-class man's mobility. The imaginations of speed, human-machine conjunctions and free-floating mobility are obviously kept between bodies gendered male. It is significant that issues of intersubjectivity, care and social connections are conspicuous by their absence in such representations. Also, the way in which even non-essentialist identities are always boundary projects, emerging from double processes of identification and distinction, tend to vanish in the metaphors of mobility (Pratt 1999).

The second objection is that more often than not, it is uncertain what the notion of mobility shall actually count for. A whole range of meanings seem to be at stake; stretching from travelling of different kinds and for different purposes, all kinds of bodily movement, processes of change and technological development, identities and new figurations of the self, to a combination of it all in an ontology of flows, networks, absence and acceleration (see e.g. Thrift 1996a). This indeterminacy appears troublesome to me because it supports an inclination of the mobility thesis, like many other overarching (post)modernisation theses, to be overdrawn and overdone. Mobility becomes a cultural hypothesis with dangerous elements of exaggeration expressed in homogenising and simplifying narratives. Part of that is an inclination towards technological determinism, unproblematically reading off characteristic of the technologies involved onto the social and cultural field.

Finally, and in continuation of the two other reservations, the understanding of the time-space configurations involved in these accounts seem one-directed and univocal. The narratives of flows, speed and acceleration are ones of the ultimate abstraction of time and space (cf. Lefebvre 1991). They tell the story of times and spaces emptied of meaning and content, consumed in ceaseless movement and acceleration. These accounts then work by suppressing other urban spatialities and temporalities – as for instance practiced place or cyclical time (see also Crang 2001). That is why I prefer, instead of mobility, to approach the city through the lens of spatiality and temporality. That, of course, does not exclude accounts of mobility, for what are mobility if not conjunctions of spatiality and temporality? But it renders possible an idea of the urban not as an one-directed abstraction of time-space, but as a site where multiple spatialities and temporalities collide. The city contains living and moving bodies, but they are not bodies moving through space-time, they are performing it and making it. It is this construction of the city as a social and mental form – through simultaneity, gathering, convergence and encounters (Lefebvre 1968), and through the emotions created in these encounters (Robins 1995) – I want to discuss in the rest of this chapter.

One way to approach this theme might be through the perspective of 'rhythms of the city'. That is, to take inspiration from Lefebvre and his provisional ideas on 'rhythmanalysis' (1992), which is focusing on the spatio-temporal flows of living bodies and their internal and external relationships. Rhythms are tentatively defined as movements and differences in repetition, as the interweaving of concrete times, but it always also implies a relationship of time to space and place. Two accentuations can be drawn from that. First, it underlines the inseparability of space and time. Lefebvre talks about localised time and temporalised place to sustain the spatio-temporal reality of rhythms and their involvement in the production of space. Secondly, it accentuates the fluidity and networks of moving bodies without presupposing singular modes of temporality and spatiality.

In his version of rhythmanalysis, Lefebvre starts from the body and views the body-city relationship through a distinction and conjunction between two kinds of rhythms: 'rhythms of the self' and 'rhythms of the other'. The rhythms of the self are deeply inscribed rhythms, organising a time oriented towards private and intimate life,[2] while rhythms of the other are rhythms turned outward, towards the public. Lefebvre however does not suggest a binary opposition between the private and the public. On the contrary, he reminds us of the multiple transitions and imbrications between these poles: in apartments, houses, streets, squares and places; in family, kinship, neighbourhood and friendship relations; and in the city itself. The separation between private and public, then, is broken down with a starting point in the body and its senses, reaching out to the bustling world and linking the two.

In a few essays Lefebvre[3] discusses the rhythm of the city as a spatial temporality involving among other things flows of bodies, spectacles and sounds. But he also stresses a continuous conflict between tendencies towards homogeneity and diversity. That is, conflicts between rhythms imposed by political and economic centrality and the polyrhythmy of different cultures, languages and sexualities. These contradictions are connected to a distinction between two kinds of repetition or temporality already introduced by Lefebvre in his writings on everyday life; that is, cyclical time and linear time. The distinction and interaction between the cyclical and the linear manifests itself both in the body and in the city; both combine the cycles of day and night, need and desire with the linearity of gestures and manipulation of things – a measured, imposed and exterior time.

In this way, rhythmanalysis can connect two dimensions of urban culture, which is already recognized by Lewis Mumford in his classic *The City in History* (1961). The first dimension is the one that Mumford refers to as the capacity of the city as a container; it not merely hold together a larger body of people and institutions than any other kind of community, it also transmits a larger portion of their lives than human memories can transmit by words. It is about the city as a supporting and facilitating environment. The other one accentuates that urban experience is not only about dwelling, it is also about movement and mixture, about mobilisation, encounters and challenges. The city is a site of stimulation, provocation and meetings with the 'stranger'. This cultural intermixture of dwelling and movement is important, not least because urban literature to a considerable degree oscillates between the two, privileging one or the other as *the* urban culture. Each of these dimensions has its own narratives and its own spatialities and temporalities, thus colliding in the multiple rhythms of the city.

Rhythmanalysis then, with its emphasis on the interaction between multiple spatialities and temporalities, transcends the narratives of speed, acceleration and one-directed abstraction of time-space. That does not mean, of course, that processes of acceleration and abstraction should be denied. Lefebvre himself is one of the most prominent observers of such processes. It is however only one amongst different trends in the spatio-temporal organisation of urban life. Others might be traced. One, for instance, is the progressive colonisation of the night – the steady movement of social life into the dark (Thrift 1996a, Crang 2001). 'Night life' is an integral part of the urban experience, developed concurrently with lighting of streets and houses and expressed in what we could call the 'non-stop' city. All spheres of life are affected by this process; the cultural construction of urban 'dream spaces' – e.g. hotels, theatres, cafés and cinemas – the timing and forms of social interaction in households – culminating in the diffusion of television and personal computers – and the organisation of work with increasing amounts of shift work and night work. The point is that the 'night people' (whether it is night workers, amusement seekers or 'underground' groups) follow particular rhythms; they perform their own spatialities and temporalities and construct their own narratives of the city. More generally, such considerations might be extended to other changes in the spatio-temporal organisation of social/urban life – in the organisation of work, in the conventions of family life or in the emergence of new social groups and life-styles. Altogether processes that might change the beat of urban rhythms, making them less conjoined and more fragmented than the ones observed by Lefebvre from his window in Paris.

The scheme of rhythmanalysis should not be overdrawn, however. It is an incomplete project, introduced through a few concepts and hypotheses and employed in very different and diverse examples. Therefore, instead of following Lefebvre's ambitions of a general 'rhythmology', I prefer to see it as a perspective from which to elucidate 'the construction of the city'.

In order to pursue this idea, a relevant place to start from is the one of urban everyday life and people's movements in the city (in walking or in other ways). Much literature approaches that by means of Benjamin's by now overly familiar figure of the 'flanêur' – the stroller and the spectator of the modern city. Such a strategy would however import into the analysis an one-directedness similar to the one discussed above, together with a visualism and a gender-bias which, as showed by many, are inherent in the figure as well (Buck-Morss 1989, Wolff 1985, 1994). When the flanêur 'goes botanising on the asphalt', he does so as a detached spectator and his visions are mediated through a male gaze, objectifying women as part of the urban landscape. There are therefore good reasons to search for an alternative strategy, one that from the start is much more sensitive to the multiplicity of experiences achieved in urban everyday practices and to the multiplicity of spatialities and temporalities involved in these practices. One road to follow in this effort could be the one suggested by de Certeau (1998) when he is considering two 'networks' decisive elements in people's active construction of the city; *gestures* and *narratives*. These two can both be characterised as chains of *operations* done on and with the surroundings. In two distinct modes, one practical and the other linguistic, gestures and narratives manipulate objects, displace them, and modify both their distribution and their uses. Following this strategy, in the rest of the chapter I will try to develop a framework for approaching the everyday construction of the city by considering in turn what I would call 'the embodied city' and 'the narrative city' respectively.

### The embodied city/ies

In this section, I want to explore the construction of the city through the perspective of embodied practices. This approach takes its starting point from the idea that lived experience and social practice is intrinsically corporeal. One text often cited for this purpose is Grosz's (1992) interesting essay on 'Bodies-Cities' in which she points to the city as one ingredient in the social constitution of the body. That is, 'the form, structure, and norms of the city seep into and affect all the other elements that go into the constitution of corporeality and/as subjectivity' (1992:248) In spite of the intention to deal with co-constitution, however, the idea of mutually defining relations seems to vanish from the argument in favour of an installation of the city as just one more element in the social construction of the body.[4] Here, I want to forward a much more ambiguous understanding of the social body.[5]

Such a one might start from Merleau-Ponty's phenomenology (1962) which identifies the body as part of a pre-discursive social realm based on perception, practice and bodily movement. Lived experience, to him, is always and necessarily embodied, located in the 'mid-point' between mind and body, or subject and object. This means that the human body takes up a dual role as both the vehicle of perception and the object perceived, as a body-in-the-world – *a lived body* – which 'knows' itself by virtue of its involvement and active relation to this world. The 'body-subjects' are not locked into their private world, but are in a world that is shared with 'others'. Consequently, to meet or to see the other is not to have an inner representation of him/ her either, it is to be-with-him/her. This underlines the understanding of the world as a genuine human interworld and of subjectivity as publicly available; the subjects are sentient-sensuous bodies whose subjectivity assumes embodied and public forms. However, the corporeality of social practices concerns not only this sensuous, generative and creative nature of lived experiences, but also the way in which these embodied experiences themselves form a basis for social action. Bourdieu (1977,

1990) for instance, talks about 'habitus' as embodied history, which is internalised as a second nature. As a result of this, social structures and cultural schemes are *incorporated* in the agents and thus function as generative dispositions behind their schemes of action. In the case of the city, this comes close to the comment quoted from Elisabeth Grosz above.

From my point of view, the advantage of this understanding of the embodied character of social practices is that it asserts a version of social constructionism that maintains an attention to the body's material locatedness in history, practice and culture. A corresponding understanding of the body, respecting the materiality and historicity of the body as well as its gender, comes from the Norwegian feminist Toril Moi (1999).[6] She bases her understanding on a re-reading of Simone de Beauvoir, also drawing on Sartre and Merleau-Ponty. The starting point is Beauvoir's notion of the body as a '*situation*'. It is a situation amongst many others – such as class, race, nationality, biography, location etc. – but it is a fundamental situation because it forms the basis of our experiences of ourselves and of our environment. It is a situation which will always be a part of our lived experience. Following Sartre, Moi considers the situation to be a structural relationship between our projects (freedom) and the world (which includes the body). The meaning of a woman's body is connected to her projects in the world – to the way in which she uses her freedom – but it is also marked by all her other life-situations. There are countless ways of living out the specific burdens and potentialities of being a woman. The understanding of sex/gender put forward by Moi (and de Beauvoir), then, is both historical and open – it is always in a process of becoming, marked by our shifting and fluctuating experiences of ourselves in the world. The relationship between body and subjectivity is neither necessary nor accidental, it is *contingent*.

Part of the locatedness of the body referred to above is a matter of its situatedness in space and time, and this opens an avenue for an understanding of the relationship of the corporeality of practice to space-time and through that to the city. First, it can accentuate the 'place character' of space (see e.g. Casey 1993, 1997). The body is always in place; notwithstanding developments of 'placelessness', 'disembeddedness', mobility and 'hyperspace', we cannot escape that fact. More important, however, is the way in which the body itself is spatial. To explore that we can once more turn to Merleau-Ponty. Initially, he states that the spatiality of the body is not a spatiality of *position*, but one of *situation*. This goes for temporality as well, and it means that we should avoid thinking of our bodies as being *in* space or *in* time – they *inhabit* space and time:

> I am not in space and time, nor do I conceive space and time; I belong to them, my body combines with them and includes them. The scope of this inclusion is the measure of that of my existence (Merleau-Ponty 1962:140)

This means that active bodies, using their acquired schemes and habits, position their world around themselves and constitute that world as 'ready-to-hand', to use Heidegger's expression. And these are moving bodies 'measuring' space and time in their active construction of a meaningful world.[7] This also accentuates the indispensable intertwining of matter and meaning. For Merleau-Ponty, the material world is not juxtaposed with an ideational one:

Merleau-Ponty refuses to separate the ideational and the material. All ideas and meanings are necessarily embodied (in books, rituals, speech, buildings etc), he maintains, and all matter embodies meaning and derives its place in the human world by virtue of that meaning ... In this sense then he calls our attention ... to *the embodiment of culture*, and he extends his argument against the abstraction of meaning and matter (Crossley 1995:59)

Notwithstanding internal disagreements,[8] I think Lefebvre (1991) can add something to this with his stronger emphasis on social practice and the production of space. He establishes a material basis for the production of space consisting of:

a practical and fleshy body conceived of as totality complete with spatial qualities (symmetries, asymmetries) and energetic properties (discharges, economies, waste) (Lefebvre 1991:61)

An important precondition of this material production is that each living body both *is* and *has* its space; it produces itself in space at the same time as it produces that space.

Like Merleau-Ponty, Lefebvre assigns an important role to the body in the 'lived experience'. As a part of that, the body constitutes a practico-sensory realm in which space is perceived through sight, smells, tastes, touch and hearing. It produces a space which is both biomorphic and anthropological. The relationship to the environment is conducted through a double process of orientation and demarcation – practical as well as symbolic. These processes are connected in a conception of '*the spatial body*':

A body so conceived, as produced and as the production of space, is immediately subject to the determinants of that space ... the spatial body's material character derives from space, from the energy that is deployed and put to use there (Lefebvre 1991:195)

When Lefebvre refers here to the energy of the body, it is not only a material/ biological notion. With reference to Nietzsche, he emphasizes the Dionysian side of existence according to which play, struggle, art, festival, sexuality and love – in short, Eros – are themselves necessities. They are parts of the transgressive energies of the body. Further, it is important to notice, this concerns not only the material and meaningful production of space, but also the capacity to transgress the 'everydayness' of modern life. It involves participation and appropriation of space for creative, generative, bodily practices, as formulated for instance in the 'right to the city' (Lefebvre 1968).

What I have proposed so far, by the help of Merleau-Ponty and Lefebvre, is an indispensable relationship among practice, body and time-space. The connection of 'body-subjects' with space can never be one of simple location; it is an active engagement with the surrounding world involving a production of meaning. The social body can then be seen as 'the geography closest in' – as a constitutive social spatiality reaching out towards other socio-spatial scales from local, urban and regional configurations to national and supra-national/global connections. In order to specify this connection on the urban scale, I shall return to de Certeau (1984) and his account of the constructiveness of moving bodies through the metaphor of 'walking in the city'.

What de Certeau is talking about here is practitioners of the city following the urban pathways, but at the same time producing their own stories, shaped out of the fragments of trajectories and alterations of spaces. It is a process of narration in which walking in the streets mobilises other subtle, stubborn, embodied and even resistant meanings. With reference to Merleau-Ponty he writes:

> These practices of space refer to a specific form of *operations* ('ways of operating'), to 'another spatiality' (an 'anthropological', poetic and mythic experience of space), and to an *opaque and blind* mobility characteristic of the bustling city. A *migrational*, or metaphorical, city thus slips into the clear text of the planned and readable city (de Certeau 1984:93)

Walking in the city, then, is one of those everyday practices or 'ways of operating' which make up the game of ordinary people, of 'the others', when they move along in a creative and tactical way in a network of already-established forces and representations. It is a spatial practice using and performing the urban system in a way that secretly influences the determining conditions of urban life. It is a *lived space* – a space of disquieting familiarity with the city. Here, the affinity with both Merleau-Ponty and Lefebvre becomes suggestive.

With inspiration from theories of ordinary language use, walking in the city is dealt with as a speech act. The act of walking, de Certeau argues, is to the urban system what the speech act is to language or to the statements uttered. Walking is a space of 'enunciation', in this sense having a triple function: it is a process of *appropriation* of the urban topographical system; it is a spatial *realization* or acting-out of place; and it implies *relations* among differentiated positions, that is, among pragmatic 'contracts' in the form of movements (or moves in relation to someone).

Altogether, this means that walking practices cannot be seen only as simple movements, they rather spatialise. Their intertwined parts give their shape to spaces. They weave places together. In that respect, walking practices create a diversity of subsystems whose existence in some sense makes up the city. Different characteristics of use are at work in this process. While a pre-given spatial order can organise an ensemble of possibilities and interdictions, the walker actualises some of these possibilities. S/he makes them exist as well as emerge. But s/he also moves them about and invents others, since the improvisation of walking privileges, transforms or abandons spatial elements. Thus the walker creates discreteness, whether by making choices among the signifiers of the spatial 'language' or by displacing them through the use of them. In this realisation and appropriation of space, the walker also constitutes a here and a there; s/he constitutes a location and in relation to that a location of an 'other', thus establishing a conjunctive and disjunctive articulation of places. The whole 'rhetoric of walking' has a 'phatic' aspect; through contacts in meetings, followings, networks etc., it creates a fluid and mobile organicity in the city.

Even if de Certeau thus produces an understanding of urban life from 'closest in', his contribution is not a voluntaristic micro-sociology (see also Crang 2000). His moving bodies (or agents) are constituted against a monolithic vision of power – against imaginary totalisations produced by the eye. One of these is the panoptic and strategic discourse of planning and social theory which rationalises the city and organises it by speculative and classificatory operations. Another one is the imaginary discourse of media and commerce (de Certeau 1997). They create cities that are 'imaginary museums' and as such form counterpoints to cities at work. These imaginary entities are characterised by a growing eroticisation, by a celebration of the body and the senses. But it is a fragmented body, categorised by virtue of an analytical dissection, cut into successive sites of eroticisation. In this connection De Certeau even talks about speech and signification as 'denaturing' acts.

These arguments are close to the ones put forward by Lefebvre (1991) when he is linking the history of the body with the history of space and, to use Gregory's (1994) word, understanding the development of modernity and the modern city as a *decorporealisation* of space. This process involves a logic of visualisation and one of metaphorisation; living bodies, the bodies of 'users' are caught up, not only in the toils of parcelised space, but also in the work of images, signs and symbols. These bodies are transferred and emptied out via the eyes, a process that is not only abstract and visual, but also phallocratic. It is embodied in a 'masculine' will to power and, metaphorically, abstract space and its material forms symbolise force, male fertility and violence.

As we have seen, however, the moving body serves both authors as a critical figure as well. It is not possible totally to reduce the body or the practico-sensory realm to abstract space. The body takes its revenge – or at least calls for revenge – for example in leisure space. It seeks to make itself known, to gain recognition as '*generative*', thus appropriating or shaping urban space through everyday life, struggle (Lefebvre) or tactics (de Certeau). One problem in these formulations is the flavour of authenticity they tend to ascribe to the body and everyday life. By using the conception of the social body put forward above, I do however think that it is possible to avoid romanticising the body and to understand the role of moving bodies in the constitution of and conflict over urban space as a historical, open-ended process.

By the help of these accounts, then, it is possible to gain a conception of the embodied city as a ceaseless spatio-temporal process of construction. What they do not provide to a satisfactory degree, however, is an acknowledgement of the differences among bodies – that is, how the cultural embodiment of the city is gendered and marked by other bodily ascribe identities.[9] Earlier in this section, I indirectly introduced this issue with the aid of Moi's/de Beauvoir's notion of the body as a situation. As a preliminary attempt to spatialise this notion, it is worth considering an original contribution from Iris Marion Young (1990) in which she explores the possibility of a specifically 'feminine' body comportment and relation in space. She displays a contradictory spatiality primarily based on the historical and cultural fact that women live their bodies simultaneously as subjects and objects. A woman in our culture experiences her body on the one hand as background and means for her projects in life. On the other hand she lives with the ever-present possibility of being gazed upon as a potential object of others' intentions. This ambiguous bodily existence tends to 'keep her in her place', and it influences her manner of movement, her relationship to her surroundings and her appropriation of space. Probably other deviations from the 'neutral' body – such as skin colour, age and sexuality – can in similar ways give rise to specific practical and symbolic spatialities.

What does that mean, more specifically, for the gendered construction of the city? Young's argument implies that feminine spatiality involves not only an experience of spatial constitution, but also one of being 'positioned' in space. That is why feminine existence tends to posit an enclosure between itself and the space surrounding it, such that the space belonging to it is constricted but is also a defence against bodily invasion. This is about power relations and in the last instance about fear of violence. For women being 'wise' is about not frequenting the city at specific times or in specific places, not wearing specific clothes etc. On the basis of her studies in Finland, Koskela (1997), however, stresses the importance of analysing courage as well as fear. She argues that women not only passively experience space but also actively take part in its production. Some women's confidence about going out produces space that is available to other women. Walking in the street can be seen as a political act: women 'write themselves onto the street' (1997:316)

These formulations open up an entry into two broader discussions of relevance to the embodied city. The first one is about the gendered (and sexualised) character of the binary opposition between public and private space, one that has been central in both feminist geography and feminism in general. It is now a well-known story how the female body has culturally and historically been associated with private space and restricted in public space,[10] and how this distinction has been reproduced in planning and urban design.[11] What is also important, however, is to recognise how this distinction is also continually (re)negotiated, through collective actions, through dress and body language and through just 'being there' (see e.g. Duncan 1996, Valentine 1996, Koskela 1997).[12]

The second discussion concerns the role of emotions in the construction of the city. Body-subjects do have emotions, and as it is argued by Robins (1995), it is necessary to acknowledge the significance of passions – both creative and destructive – in urban culture. For instance, the embodied city involves encounters with different bodies or 'strangers' and the emotions such meetings create, whether these are enjoyment and desire or rather anxiety, fear and aggression (Robins 1995).

## The narrative city/ies

> We dream in narrative, day-dream in narrative, remember, anticipate, hope, despair, believe, doubt, plan, revise, criticize, construct, gossip, learn, hate and love by narrative. In order really to live, we make up stories about ourselves and others, about the personal as well as the social past and future (Barbara Hardy, here from Finnegan 1998:1)

This comment well sums up how we organise our experience and our memory of human happening and social and material surroundings by the way of narratives – stories, excuses, myths, reasons for doing and not doing, and so on. The acknowledgement and treatment of the important role of narrative in organising our knowledge and our experience is of course not a new phenomenon. There has been a long tradition in literary studies, socio-linguistics, anthropology and folklore focusing on the arts, structures and functions of stories. But recent years have experienced an explosion of interest in the constructive role of narrative. The scope has extended from the 'traditional' contexts into studies right across the human and social sciences. This, as well, has given occasion for a stressing of the social character of narrative. As for instance formulated by Mumby (1993), narrative should be seen as a socially symbolic act in the double sense that (a) it takes on meaning only in a social context and (b) it plays a role in the construction of that social context as a site of meaning within which social actors are implicated.

What is a narrative then, and how does it connect to the city? For the purpose of this chapter I shall not enter into the vast and conflict-full literature on definitions, but just outline some of the features that I consider significant for the focus on the narrative 'construction' of the city. As a starting point, and in relation to the above, I am interested in narratives from the point of view of performed tellings rather than decontextualised texts with purely cognitive import. They are presentations of events or experiences which are *told* (through written or spoken words), and their account involves questions of context, delivery and active participants. Basically, such an approach can seek support in the classic text of Ricoeur (1984) in which he, with reference to Aristotle's *Poetics*, analyses narrative through the conceptual pair 'muthos-mimesis'. These two refer to the constitutive act of 'emplotment' – the organisation of events into some kind of 'whole' – and the practice of 'representation', respectively. And, as it is strongly emphasised by Ricoeur, the approach through the 'poetic' lends a sense of construction and dynamism to this pair of terms, which subsequently have to be taken, not as structures, but as operations. Let me slightly elaborate on that by the help of a few additional properties of narratives – selectively drawn from Bruner (1991) and Finnegan (1998). The first one is their *particularity*. Narratives take as their ostensive reference particular social happenings and/or environments. But this is generally their vehicle rather than their destination. The particularity is mostly embedded in more general scripts or generic conventions – that is, something of the universal in the particular. Secondly, narratives have in their organisation some element of explanation or *coherence*. That is what 'emplotment' is generally about – some kind of organisation of events that involves a sense of intelligible order and/or 'causality'. Thirdly, I want to bring up the *context sensitivity* and negotiability of narratives. This is a quality that connects to both the telling and the interpretation of narratives and relies basically on accounts of narrative intention and background knowledge. It is this very context sensitivity that makes narrative discourse in everyday life such a viable instrument for cultural negotiation (Bruner 1991). Finally, one feature turns up in nearly all accounts of narrativity – one that needs further elaboration because of its importance to the narrative construction of the city. That is narrative diachronicity or the *temporality* of narrative.

One way of approaching the relationship between time and narrative is through the lens of the phenomenology of time. Time in these writings is considered constitutive of human being-in-the-world and, more important in the present connection, it is seen as a structural connection of future, past and present (Heidegger 1962, Ricoeur 1984). This connection takes form of a 'threefold' present (the present of the future, the present of the past, and the present of the present) indicating that a person's present disposition and actions can only be made sense of in relation to a future and a past. Furthermore, these three dimensions are seen as moments of mental action – as acts of expectation, memory and attention, respectively – each considered not in isolation but in interaction with one another (Ricoeur 1984).[13]

Already the introduction of these terms – expectation, memory and attention – suggest a connection between time and narrative, but Ricoeur takes us further than that. He adds a (maybe too)[14] far-reaching thesis on a correlation between the activity of narrating a story and the temporal character of human experience seeing it as not merely accidental but presenting a transcultural form of necessity, and he does so by considering three stages of representation or the telling of a story. The first stage is about refiguration or the way in which stories are grounded in a preunderstanding of the world of action, its meaningful structures, its symbolic resources and its temporal character. Any narrative understanding is in some way connected to practical understanding, to the familiarity of ideas such as agent, motive, means, cooperation, conflict, success, failure etc. – or in other words, narratives inevitably draws from an explicit or implicit phenomenology of 'doing something'.[15] This practical field in return opens itself to narration, because symbolic resources such as signs, rules and norms, always already articulate it, and because it is organised in temporal structures that call for narration. For the sake of the last point, Heidegger's thinking on the temporality of Being is drawn on to argue for a prenarrative quality of experience, an understanding of life as '(as yet) untold' stories (1984:74). That is, episodic events and repressed stories waiting to be collected and retold as life stories and personal identities. The second mode of representation concerns the creative operation of configuration or emplotment. Plot in this process exercises a mediating function; by drawing a meaningful story from a diversity of events or incidents, by connecting factors as heterogeneous as agents, goals, means, interactions, circumstances and unexpected results, and by temporal ordering of episodes into 'series', 'successions' and 'endings'. Altogether constituting the productive imagination of narrative. Finally, representation also involves rhetoric and 'reading', that is, the intersection of the world of the text and that of the listener or the reader. This stage is about the application of narrative, a dialogical process that most precisely can be described by Gadamer's notion of the 'fusion of horizons'. By this threefold conception of representation, Ricoeur on the one hand retains a connection between narrative and the world of practice and experience, on the other he recognises indefiniteness in this connection, one that is approached (but never remedied) through emplotment or the poetics of narrative. Also, his passage through Heidegger's phenomenology of time has put attention to the hierarchisation of the levels of temporality and to the passage from 'authentic' or mortal time towards everyday time and more public and multiple temporalities.

It is exactly such multiple temporalities that are at stake when we consider the narrative construction of the city. Memories and expectations on the part of different urban agents constitute performative speech acts in the production of urban imaginaries – from autobiographical remembering to artistic description, planning and urban marketing. Nostalgia and the imaginary city of the past are fused in public discourses for example in images of the community or heritage industries, just as futuristic visions of speed and cyborg-capitalism are part of the urban imaginary. In accordance with the insight gained from the phenomenology of time, however, the temporality of narrative is not a question of a past of memory, a present of description and a future of imagination and planning:

> The past exists as the projection backwards of present concerns. The desire for a good city in the future already exists in the imagination of the past. The future tense of urbanist discourse turns out to be less predictive than optative, although expressed in the present tense of the architectural drawing (Donald 1997:184)

Urban narratives then – whatever told by city-dwellers, artists, planners, urban theorists or others – should be seen in the light of this fusion of past, future and present. Having said that, however, it is obvious that such narratives are structured by different temporalities and different temporal orderings – from the narratives of city-dwellers structured by everyday lives and life cycles (including markable events and turning points) to 'grand' narratives of urban theorists ordered by more or less evolutionary historical periodisations (Simonsen 1993, Finnegan 1998).

While the bulk of literature on narrative and narrative inquiry emphasizes the diachronicity and the temporality of narratives, questions on the spatiality of narrative call upon other sources, some of which have been widely discussed in geography during the latest decades. For my purpose, an expedient starting point is the contribution from Michel de Certeau (1984) who emphasizes the relationship between practices and narratives as well as the spatiality of both. To him, stories are spatial trajectories:

> Every story is a travel story – a spatial practice. For this reason, spatial practices concern everyday tactics, are part of them, from the alphabet of spatial indication ('it's to the right', 'take a left'), the beginning of a story the rest of which is written by footsteps, to the daily 'news' ('guess who I met at the bakery?'), television news reports ('Teheran: Khomeini is becoming increasingly isolated'), legends (Cinderellas living in hovels), and stories that are told (memories and fiction of foreign lands or more or less distant times in the past). These narrated adventures, simultaneously producing geographies of actions and drifting into the commonplaces of an order, do not merely constitute a 'supplement' to pedestrian enunciation and rhetorics. They are not satisfied with displacing the latter and transposing them into the field of language. In reality, they organize walks. They make the journey, before or during the time the feet perform it (1984:115–116)

What de Certeau is arguing in this comment is that narrative is conditioning a map – a 'cognitive map' (cf Jameson 1991) – but one that is less inclined to produce an overall vision than to represent everyday life. It is composed by a range of active narrative operations or speech acts and always related to spatial practices. Let me in the following shortly consider the function of such 'spatialising' narrative actions.

*Spaces and places*

Stories, in de Certeau's words, carry out the labour of transforming 'places into spaces and spaces into places' (1984:118). As spatial trajectories, they every day traverse and organise places; they select them and link them together; they make sentences and itineraries out of them. De Certeau identifies two poles of description or narrative figures in this process – that is the 'tour' and the 'map'. In systemic discourse, these two over the past five centuries have been slowly dissociated in literary and scientific representations of space. In everyday narratives, however, they coexist and interlace. Narrative oscillates between them, combining the experience of moving (spatialising actions) and seeing (the knowledges of an order of places). In this way stories organise the play of changing relationships between spaces and places, but also in a pre-established geography they tell us what we can do in it and make out of it. Stories however not only traverse and organise places they also make them habitable. As well-known from any notions of sense of place, memories, legends and other signifying practices can saturate places with meaning and emotions, even though these refer to lack as often as presence, fragmentation as often as coherence.

*Boundaries and bridges*

Just as we saw in the case of traversing and organising places, stories also have an important role in cases concerning their delimitation. This is of course most evident when it is made explicit as in juridical discourse, but the role is a much more general one. De Certeau describes its primary function as one of authorising the establishment, displacement and transcendence of limits. This means that it, in the field of discourse, sets in opposition two movements that intersect – setting and transgressing limits – and constitutes a dynamic partitioning of space by the way of the two narrative figures 'boundaries' and 'bridges'. Hillis Miller (1995) conducts a similar argument. He draws on Heidegger's notions of 'Riss', 'Brücke' and 'Ring' to illustrate how narratives ground themselves in a landscape. The point is that the associated speech acts – dividing, conjoining, encircling – do not set limits *in* space they rather produce space. They project events onto space and by that produce a narrational space. By the way of these narrative figures, then, stories are founding or creating a field (although fragmented and polyvalent) for practical actions.

Boundaries and bridges are however ambiguous narrative categories which constitutes a complex relationship of contradiction and connectivity:

> In the story, the frontier functions as a third element … A middle place, composed of interactions and interviews, the frontier is a sort of void, a narrative symbol of exchanges and encounters (de Certeau 1984:127)

The boundary establishes a space 'in-between' to see through, just like the bridge alternately welds together and opposes insularities. The two of them presupposes each other and in complex ways both distinguishes and connects. Any narrative of delimitation and enclosure in some sense presupposes a relation to an alien exteriority; it is as if delimitation itself was the bridge that opens the inside to its other.

*Territoriality, othering and naming*

The ambiguity and the emotions involved in the construction of boundaries and bridges become even more important when it comes to the now widespread discussion of power in narrative spatialisation. The importance of narrative in the construction of of territory has been widely acknowledged since Anderson (1993) argued that nation states are 'imagined communities', constructed through a range of media and connected to the upcoming of print capitalism. On the urban scale, Cohen (1997) has demonstrated such mechanisms in his 'narrative cartography' of the white population of London's Isle of Dogs. He explores the popular imagery of the community through the 'simple' equation of story, plot, territory and map and identifies an increasingly racialised articulation of the East End story. Spatialised othering is however not confined to the narrative construction of territory. Rather, this should be seen as a special case of the more extensive process of constructing imaginative geographies. The understanding of that more than anything else owes to postcolonial discourse, in particular to Said's *Orientalism* that so convincingly put the violence of representation on the agenda (see, among many others, Gregory 1994, 2000, Slater 1999). One element in such power strategies is naming. For colonisation the practice of naming was a way of bringing the landscape into textual presence, of bringing it within the compass of a European rationality that made it at once familiar to its colonisers and alien to its native inhabitants (Gregory 1994). As this suggest, at the same time as being a power strategy, naming is also a contested and ambiguous area. In spaces brutally lit to by an alien reason, de Certeau argues, proper names carve out pockets of hidden and familiar meaning. They 'make sense', construct alternative places and passages, and as such ascribe meaning to practices of evasion, undermining and resistance (see also Pred 1990). Also, names can assume mythical form partially detached form the places they were originally supposed to define, like when names of streets, squares and buildings make themselves available to the diverse meanings given them be passers-by and serve as metaphor or meeting points on imaginary itineraries. Altogether articulating a second, poetic geography on the top of the literal, forbidden or permitted meaning.

Narrative is not the whole life. But it is no small part of our social existence that we can create storied pathways to live by, including performance conventions, plots, order, myths etc. Exploring such narrative resources for creating and experiencing our surroundings, therefore, can institute an understanding of the city as a collection of stories. De Certeau preliminarily describes the relation between spatial and signifying practices by way of the three symbolic mechanisms of legend, memory and dream. They make places habitable and believable, they recall or suggest phantoms and they organise the invisible meanings of the city. The temporality and spatiality of narrative, as it is discussed above, can elaborate this relationship. They establish how memories and expectations intersect in present dispositions for action and the constitution of urban imageries; how the poetic act of emplotment involves temporal and spatial 'ordering' of events; how narrative figures such as tours, maps, boundaries and bridges are integral parts of the construction of social life and, in particular, the organisation of spaces and places; and finally how narrative power strategies, involved in processes of territoriality and othering, constitute hierarchies of places at different scales – including the urban one.

'The narrative city' then is constituted by a *multiplicity* of stories of urban life, told by a range of tellers creating different social, temporal and spatial contexts. The urban agents are active tellers with the capacity to draw from continually recreated narrative resources and narrative figures. Some of the stories in question might be widely agreed among different groups of people, others are told in specific contexts and for particular purposes. Some may recount rather general characteristics of the city, others tell of specific parts of the city or of the 'unique' individuals who people them. Many different kinds of stories are told about the city. In this paper, most emphasis is put on stories referring to the everyday life of city-dwellers, but they of course intersect with other narratives. For example, everyday accounts of the city are influenced by the myriad of imaginative cityscapes brought to us by novels, films and other media, also – and connected to that – narrations generated by practices of urban planning, local politics or city entrepreneurs are discourses interfering in everyday life.

These interrelationships makes topical what discourse analysis has learned us so well, that narrative constructions are imbued with power constituting contradictions, hegemonies and narrative hierarchies. My concern in this place, however, is not a search for dominant discourses or 'nodal points' (Laclau and Mouffe 1985), but for the multiplicity of stories constituting the city not as one but as *many* cities. The emphasis on the multiplicity of narratives, referring to a whole range of different urban groupings and practices, should also be seen as an attempt to overcome the tendency to dichotomous thinking involved in vocabularies such as power/resistance, representations of space/spaces of representation (Lefebvre) or strategies/tactics (de Certeau). The disadvantage of such dichotomies (regardless of their attractiveness) is that narratives and practices, which cannot be understood in this schematic outline, often remain hidden.

## Concluding comments

In the discussion of the construction of the city conducted in this chapter I have been trying to outline a time-space that includes *embodiment, narrativity* and *difference*. I opened the discussion by dissociating myself from ideas of the urban as a one-directed abstraction of time-space, instead seeing it as a site where multiple spatialities and temporalities collide. Thinking in rhythms of the city opens up a path to an understanding of urban life through alternating successions of interactive bodies during day and night. It starts from the body-subjects, describing the dynamic spatialities and temporalities involved in the practices of creative and moving bodies and the collision of these rhythms with those of abstraction in the construction of the city. Thus constituting what I have called the embodied city. These processes of construction however also involves narratives or storied pathways to live by; narratives that have their own temporalities (memories, expectations, temporal ordering) and their own spatialities (places, pathways, territories). That is, encompassing the narrative as well as the embodied city. It is obvious from the above that a close connection exists between the 'narrative' and the 'embodied' city. One however is not reducible to the other. The lived experience of non-verbal space is only partially translatable into language or narratable in story. In it we find the spatialisation of passions and pulsations – Eros and Thanatos – that cannot be the web of any structure or story (Mazzoleni 1993). An understanding of the construction of the city therefore involves analyses of the way in which co-constituting bodily and narrative operations give shape to urban spaces. A paradigmatic dimension of such an understanding is difference – in culture, ethnicity, class, gender and sexuality – that is, experiences and encounters between different bodies in lived, perceived and conceived space.

## Notes

1   This is the title of a research project on urban practices and strategies which I together with Esben Holm Nielsen conduct at Roskilde University.
2   Examples could be breathing, heartbeat, sexuality, thought and social life.
3   Some of them written together with Catherine Regulier.
4   For the sake of fairness, it should be made clear that this comment only address the essay in discussion here. Grosz's more general accounts of the body (e.g. 1994) are rich in nuances on the constitution of body, gender and subjectivity.

5  For a more developed version of the following discussion on the body, see Simonsen 2001a and 2003.
6  Toril Moi formulates her essay as a contribution to the sex/gender debate and partially in opposition to the Foucault-inspired notion given by Judith Butler (1990). In that, she argues, the body is reduced to a *tabula rasa* awaiting inscription from culture.
7  This idea of moving bodies is used in an interesting way used by Parviainen (1998) to develop a theory of the art of dance.
8  For instance, Lefebvre criticised Merleau-Ponty for his focus on perception and for not taking social practice sufficiently into consideration, see e.g. Lefebvre 1991:183n. For a more thorough discussion of Lefebvre's treatment of the body see K. Simonsen (2001b) 'Bodies, sensations, space and time. The contribution from Henri Lefebvre' (forthcoming).
9  Merleau-Ponty has been criticised for offering a 'neutral' phenomenology of the body building on the male body (Irigaray 1984, Young 1990) and a similar critique might be raised on de Certeau's formulations. Lefebvre stresses the 'right to difference', but does in fact not connect this to the spatial body.
10  Even if this point is not as unequivocal as sometimes suggested (see e.g. Wilson 1991).
11  For a more detailed Scandinavian discussion of this, see Simonsen (1990).
12  Part of this can be to identify public/private interspaces that condense such negotiations. Scott Sørensen (1998) and Lyngfelt (1998) identify the literary salon of the nineteenth century as such a public/private interspace and Jónasdóttir and von der Fehr (1998) suggest the workplace as a late modern one.
13  It is interesting that Ricoeur, in a passing comment, acknowledges a relation to space as well: 'How can we measure expectation and memory without taking support from the "points of reference" marking out the space traversed by a moving body, hence without taking into consideration the physical change that produces the trajectory of the moving body in space?' (1984:21).
14  Irrespectively of the degree of acceptance of this thesis, however, the argument adds valuable insight into the connection and to the character of narrative.
15  Ricoeur in this place refers to English language philosophy, for example as represented by Wittgenstein's connection of everyday practice and everyday language (Wittgenstein 1953).

## References

Anderson, B. 1993 *Imagined Communities: Reflections on the Origin and Spread of Nationalism*, London: Verso.
Bourdieu, P. 1977 *Outline of a Theory of Practice*, Cambridge: Cambridge University Press.
Bourdieu, P. 1990 *The Logic of Practice*, Cambridge: Polity Press.
Bruner, J. 1991 'The Narrative Construction of Reality', *Critical Inquiry* 18:1–22.
Buck-Morss, 1989 *The Dialectics of Seeing. Walter Benjamin and the Arcades Project*, Cambridge, MA: MIT Press.
Butler, J. 1990 *Gender Trouble: Feminism and the Subversion of Identity*, New York: Routledge.
Casey, E.S. 1993 *Getting back into Place*, Bloomimgton/Indianopolis: Indiana University Press.
Casey, E.S. 1997 *The Fate of Place. A Philosophical History*, Berkeley/Los Angeles/London: University of California Press.
de Certeau, M. 1984 *The Practice of Everyday Life*, Berkeley: University of California Press.
de Certeau, M. 1997 *Culture in the Plural*, Minneapolis/New York: University of Minnesota Press.

de Certeau, M. 1998 'Ghosts in the City' in de Certeau, M., Giard, L. and Mayol, P. *The Practice of Everyday Life. Volume 2: Living and Cooking*, Minneapolis: University of Minnesota Press, 133–145.

Cohen, P. 1997 'Out of the melting pot into the fire next time: Imagining the East End as city, body, text', in Westwood, S. and Williams, J. (eds) *Imagining Cities: scripts, signs, memory*, London: Routledge.

Crang, M. 2000 'Relics, places and unwritten geographies in the work of Michel de Certeau (1925–86)', in Crang, M. and Thrift, N. (eds) *Thinking Space*, London/New York: Routledge.

Crang, M. 2001 'Rhythms of the City: Temporalised Space and Motion', in May, J. and Thrift, N. (eds) *Timespace Geographies of Temporality*, London: Routledge, 187–207.

Crossley, N. 1995 'Merleau-Ponty, the elusive body and carnal sociology', *Body & Society*, 1:1, 43–65.

Deleuze, G. and Guatterie, F. 1988 *A Thousand Plateaus. Capitalism and Schizophrenia*, London: Athlone Press.

Donald, J. 1997 'This, Here, Now. Imagining the modern city' in Westwood, S. and Williams, J. (eds) *Imagining Cities. Scripts, signs, memory*, London: Routledge.

Duncan, N. 1996 (ed) *BodySpace*, London/New York: Routledge.

Finnegan, R. 1998 *Tales of the City. A Study of Narrative and Urban Life*, Cambridge: Cambridge University Press.

Gregory, D. 1994 *Geographical Imaginations*, Oxford: Blackwell.

Gregory, D. 2000 'Edward Said's imaginative geographies', in Crang, M. and Thrift, N. (eds) *Thinking Space*, London: Routledge, 302–349.

Grosz, E. 1992 'Bodies-Cities', in Colomina, B. (ed) *Sexuality and Space*, Princeton papers of Architecture, New York: Princeton Architectural Press.

Grosz, E. 1994 *Volatile Bodies. Towards a corporeal feminism*, Bloomington/Indianapolis: Indiana University Press.

Heidegger, M. 1962 *Being and Time*, Oxford: Blackwell.

Hillis Miller, J. 1995 *Topographies*, Stanford: Stanford University Press.

Irigaray, L. 1984 *Etique de la difference sexuelle*, Paris: Minuit.

Jameson, F. 1991 *Postmodernism, or, the Cultural Logic of Late Capitalism*, Durham: Duke University Press.

Jónasdóttir, A.G. and von der Fehr, D. 1998, 'Introduction: Ambiguous Times – Contested Spaces in the Politics, Organisation and the Identities of Gender' in von der Fehr, D., Rosenbeck, B. and Jónasdóttir, A.G. (eds) *Is there a Nordic Feminism?*, London: UCL Press, 1–18.

Koskela, H. 1997 ' "Bold walk and breakings": women's spatial confidence versus fear of violence', *Genderr, Place and Culture*, 4:3, 301–319.

Laclau, E. and Mouffe, C. 1985 *Hegemony and socialist strategy: Towards a radical democratic politics*, London: Verso.

Lefebvre, H. 1968 *Le droit   la ville*, Paris:  ditions Anthropos.

Lefebvre, H. 1991 (orig.1974) *The Production of Space*, Oxford: Blackwell.

Lefebvre, H. 1992  *lements  de rythmanalyse: Introduction   la connaissance de rythmes*, Paris: Syllepse.

Lyngfelt, A. 1998 'The dream of reality: a study of feminine dramatic tradition, the one-act play' in von der Fehr, D., Rosenbeck, B. and Jónasdóttir, A.G. (eds) *Is there a Nordic feminism?* Berkeley: University of California Press.

Massey, D. 1994 *Space, Place and Gender*, Oxford: Polity Press.

Mazzoleni, D. 1993 'The City and the Imaginary', in Carter, E., Donald, J. and Squires, J. (eds) *Space and Place. Theories of Identity and Location*. London: Lawrence & Wishart.

Merleau-Ponty, M. 1962 *Phenomenology of Perception*, London: Routledge and Kegan Paul.

Moi, T. 1999 *What is a Woman?* Oxford: Oxford University Press.

Mumby, D.K. 1993 (ed) *Narrative and Social Control: Critical Perspectives*, London: Sage.

Mumford, L. 1961 *The City in History*, London: Secker and Warburg.

Parviainen, J. 1998 *Bodies moving and moved. A Phenomenological Analysis of the Dancing Subject and the Cognitive and Ethical Values of Dance Art*, Tampere: Tampere University Press.

Pratt, G. 1999 'Geographies of Identity and Difference: Marking Boundaries', in Massey, D. Allen, J. and Sarre, P. (eds) *Human Geography Today*, Cambridge: Polity Press.

Pred, A. 1990 *Lost words and lost worlds: Modernity and the language of everyday life in late nineteenth century Stockholm*, Cambridge: Cambridge University Press.

Ricoeur, P. 1984 *Time and Narrative* vol 1, Chicago: University of Chicago Press.

Robins, K. 1995 'Collective Emotion and Urban Culture' in Healy, P., Cameron, S., Davoudi, S., Graham, S. and Madani-Pour, A. (eds) *Managing Cities. The New Urban Context*, Chichester: John Wiley & Sons.

Scott Sørensen, A. 1998 'Taste, manners and attitudes – the bel esprit and literary salon in the Nordic countries c.1800', in von der Fehr, D., Rosenbeck, C. and Jónasdottir, A.G. (eds) *Is there a Nordic Feminism?* Berkeley: University of California Press.

Simmel, G. 1969 (orig. 1903) 'Metropolis and Mental Life' in Sennett, R. (ed) *Classic essays on the culture of cities*, Englewood Cliffs, New Jersey: Prentice-Hall.

Simonsen, K. 1990 'Urban Division of space – A gender Category?', *Scandinavian Housing and Planning Research* 7: 143–153.

Simonsen, K. 1993 *Byteori og Hverdagspraksis*, København: Akademisk Forlag.

Simonsen, K. 2001a 'Rum, sted, krop og køn – dimensioner af en geografi om social praksis', in Simonsen, K. (red) *Praksis, rum og mobilitet*, Roskilde Universitetsforlag.

Simonsen, K. 2001b 'Bodies, sensations, space and time: The contribution from Henri Lefebvre', paper.

Simonsen, K. 2003 'The embodied city – from bodily practice to urban life', in Öhman, J. and Simonsen, K. (eds) *Voices from the North – New Trends in Nordic Human Geography*, Aldershot: Ashgate.

Slater, D. 1999 'Situating Geopolitical Representations: Inside/Outside and the Power of Imperial Interventions', in Massey, D., Allen, J. and Sarre, P. *Human Geography Today*, Cambridge: Polity Press, 62–85.

Thrift, N. 1996 *Spatial Formations*, London: Sage.

Thrift, N. 1996a 'Inhuman Geographies: Landscapes of Speed, Light and Power' in Thrift, N. *Spatial Formations*, London: Sage, 256–311.

Valentine, G. 1996 '(Re)negotiating the "heterosexual street"', in Duncan, N. (ed) *BodySpace*, New York/London: Routledge.

Virilio, P. 1983 *Pure War*, New York: Semiotext(e).

Virilio, P. 1991 *The Aesthetics of Disappearance*, New York: Semiotext(e).

Wittgenstein, L. 1953 *Philosophical Investigation*, Oxford: Blackwell.

Wolff, J. 1985 'The invisible flâneuse: Women and the literature of modernity', *Theory, Culture and Society* 2:3, 37–48.

Wolff, J. 1993 'On the road again: metaphors of travel in cultural criticism', *Cultural Studies*, 7, 224–239.

Wolff, J. 1994 'The artist and the flâneur: Rodin, Rilke and Gwen John in Paris', in Tester (ed) *The Flâneur*, London/NewYork: Routledge, 111–138.

Young, I.M. 1990 *Throwing like a girl and other essays in feminist philosophy and social theory*, Bloomington: Indiana University Press.

## Chapter 4

# 'Space Oddity': A Thought Experiment on European Cross-Border Mobility

Anke Strüver

> *Ground Control to Major Tom // Ground Control to Major Tom,*
> *Take your protein pills and put your helmet on.*
> *Ground Control to Major Tom // Commencing countdown, engines on.*
> *Check ignition and may God's love be with you.*
> *Ten, Nine, Eight, Seven, Six, Five,*
> *Four, Three, Two, One, Liftoff.*
> *This is Ground Control to Major Tom, you've really made the grade.*
> *And the papers want to know whose shirts you wear. Now it's time to leave the*
> *capsule if you dare.*
> *"This is Major Tom to Ground Control, I'm stepping through the door,*
> *and I'm floating in a most peculiar way and the stars look very different today.*
> *For here, am I sitting in a tin can, far above the world.*
> *Planet Earth is blue and there's nothing I can do.*
> *Through I'm past one hundred thousand miles, I'm feeling very still.*
> *And I think my spaceship knows which way to go.*
> *Tell my wife I love her very much she knows."*
> *Ground Control to Major Tom // Your circuit's dead, there's something wrong.*
> *Can you hear me Major Tom? Can you hear me Major Tom?*
> *Can you hear me Major Tom? Can you ...*
> *"Here am I floating round my tin can, far above the Moon.*
> *Planet Earth is blue. And there's nothing I can do."*[1]

'Borderless Europe', 'Europe without frontiers' and 'freedom of movement' as catchphrases and geopolitical discourses came into existence with the realisation of the European Single Internal Market. With its opening at January 1st, 1993, all barriers to trade, investment and mobility within the European Union (EU) were said to be gone or at least formally removed. This transformation was accompanied by the establishment of the 'Committee of the Regions', the idea of a 'Europe of regions' and the completion of the Schengen Treaty which established internal and external EU-borders and opened the internal ones for EU-citizens. Europeanisation is therefore a process in which state-borders are transformed into administrative boundaries, accompanied by an active promotion of cross-border regions. Former in-between zones where two (or more) national peripheral borderland areas adjoin are supposed to become European core regions through interaction across borders and unimpeded because borderless mobility. But so far, this cross-border regionalisation from above is largely a bureaucratic matter, rather than one in which the people are involved actively.

### Ten, Nine: 'Floating far above the world' – EU-phoria

This contribution confronts Brussels' view upon the European Union, its cross-border regions and their inhabitants – both literal and metaphorical a view from above – with people's everyday practices along and across inner European borders. It reveals the former as what de Certeau has called a 'visual simulacrum' that is doomed to misunderstand the latter, people's everyday practices, because of their absence from geographical space of panoptical constructions (de Certeau 1984:93). Using a 'remix' of Bowie's lyrics to structure my arguments, I will at first summarise some findings on cross-border immobility – about the non-practice of border crossing, taking the Dutch-German borderland as example – and then introduce the idea of a 'cognitive-imaginative border'. Subsequently, I will deal with the accepted significance of representations and imaginations for the spatial practices of everyday life. This includes an application of the notion of 'spaces as representations' to borders in general and to 'embodied practices' in particular in order to link socio-spatial practices with images and imaginations. Finally, I will try to re-imagine the 'odd' European discourses on borderless mobility and 'odd spaces of immobility'.

### Eight, Seven: 'The circuit is dead, there is something wrong' – The non-practice of border crossing

The European Union's attempt to create a common space that is both open and public by pulling down the borders between the member states and establishing cross-border regions (euregios), among other things, is a classic example of integration. Yet, a united Europe and a successful transformation of cross-border regions into 'regions without borders' seem to be far from being practised or even realisable. Scott (2000; 2002), for example, argues that the cross-border regions in the EU are highly structured political projects with only small achievements. He indicates provocatively that the institutionalisation of these regions is rather a way to acquire EU-funding, than to create transborder interaction and that the idea(l) of a united Europe is too far away from its citizens. Kramsch (2002), looking at Dutch-German cross-border regions, also emphasises that many people are not aware of their 'privileged position' – of living in an euregio and being expected to cross the border regularly as the most natural thing in the world. Similar results were found by Van der Velde (2000a), who has summarised recent findings on the perception of European borders and their effects in the European Single Market era. The removal of borders as barriers to integration turns out to be more difficult than expected, especially because of their persistence in people's minds. That is to say that the perceptions of the border and the beyond, the other side of the border and the people 'over there', obstruct cross-border interaction. Moreover, he argues that the European integration process is a top-down one in which regional politicians (and to a certain degree also 'the public') are involved, but the latter rather as the receiving part and the former as merely financial administrators.

Since the changes due to the territorial re-organisation as part of the European unification were expected to be very distinct in border regions, a lot has been written about the boundaries' transformation from barriers to contact zones. In fact, cross-border co-operation and mobility are two of the main EU-objectives. Borders as hindrances to interaction should be overcome. However, the social and cultural impacts of borders were acknowledged more recently, and, in turn, their effect as barriers to socio-cultural exchange and feelings of togetherness. Hence, despite the idea of a 'borderless Europe', progressing European integration and the formal establishment of cross-border co-operation, the European Union is still full of borders: The perceptions of borders and their contiguous regions as well as their persistence in people's minds obstruct cross-border interaction and thus, a border's affective and cognitive meanings seem to be thresholds in people's everyday life. Still, the opening up of borders within the European Union is said to result in an increasing permeability of borders or even borderlessness. And what is more, EU citizens are not only permitted, but invited to cross them regularly as migrants, commuters or day-trippers. But after the first decade of excitement and EU-phoria, one can observe an atmosphere of disillusionment in the meantime with respect to borderless spaces and mobility.

The process of European integration does have a stimulating impact for cross-border regions – especially economically and politically. The removal of borders as barriers to capital's, good's and people's mobility is expected to lead to equalisation. But there are socio-cultural obstacles to the free movement of people and these factors are too often marginalised. Up to now, the story of European integration can (still) be summed up as an attempt to create a common economic space. It is a story about 'political unity', 'economic growth' and 'cohesion', but not about people.

Therefore, I agree with Anssi Paasi (2001), who argues that discourses on the 'Europe of regions' are relatively separate from people's everyday lives. There is little in official cross-border co-operation that local 'borderlanders' are aware of, with which they can identify with and want to participate in. On the one hand, this is a consequence of the EU-bureaucrats' practice of ignoring people's interests, to disregard the local people in border regions. On the other hand, for people who do not know about the European Union's efforts to create cross-border regions, as well as for those who inhabit one of these borderland areas, they do not respond to cross-border initiatives and their offers. Yet, aside from this 'passive ignorance', there are many other factors that make inner European borders anything, but insignificant.

Keeping this 'officially increasing Europeanisation' in mind, I will now shift the attention to one of the pioneers for cross-border co-operation in Europe, to the 'increasing' cross-border activities between Germany and the Netherlands – a borderland where the first European cross-border association was established (in 1958, Euregio Gronau). But despite the fact that co-operation across the Dutch-German border has a long tradition within post-war Europe and a more recent, but yet intensified institutionalisation, the extent and intensity of current cross-border interaction is not remarkable yet. It is also a top-down cross-border regionalisation of which the people are not really part.

One could argue easily that there are obvious obstacles to unimpeded cross-border activities between Germany and the Netherlands, such as insufficient information, traffic conditions and cross-border infrastructure, as well as language differences – all of them 'material obstructions' in a broader sense. On the other hand, there are also various attempts to build 'material bridges' – related to the public services sector, i.e., cross border co-operation of fire brigades, police interventions, medical care etc. However, all these examples are not what I am talking about when I refer to the borders in people's minds, to the 'non-practice of border crossing'. For, these phenomena do not indicate that people do not cross the border! Both Germans and Dutch cross more or less on a regular basis for exceptional and exciting events such as tulip festivals and beach holidays or wine tasting sessions and mountaineering respectively. People also cross (to a certain extent) for shopping (Van der Velde 2000b). Eventually, they even move to the 'other side' because of huge differences in real estate prices. But these new suburbs that are built tend to be rather enclaves, 'colonies' or 'ghettos' than ordinary and fully integrated residential areas (Dortmans 2002; Rengers 2002) – and in general, an enclave is a place that does not only respect, but depends on the (maintenance of the) border!

What I mean with *non-practice of border crossing* refers to the multiple and maintained affective borders in people's minds, to cognitive-imaginative borders. The 'non-practice' is used to describe people's everyday practices that 'stop' or 'end' at the border, i.e. the border demarcates the 'spheres' of people's spatial practices. And this border is thus less effective as line or political manifestation than important as 'barrier' in people's minds.

Taking work as one, but very dominant part of everyday lives and practices, empirical analyses of the Dutch-German cross-border labour market reveals that cross-border labour mobility remains at a very low level. Apart from the border as (assumed) obstacle, this is particularly striking since a vivid and integrated cross-border labour market seems very likely due to the unemployment and wage disparities between the two national labour markets (Busse and Frietman, 1998; van Dijk and Zanen, 2000; Janssen, 1999, 2000; van der Velde, 1999). Moreover, research on young Dutch and German people shows that the willingness to *any kind of cross-border activities* is very poor (Janssen, Spille and Baerveldt 1996; see also de Bois-Reymond 1998; more generally, see Lademacher 2000; Verheyen 1993). And regarding shopping daily goods for example, another 'true' everyday practice that is not much reflected upon by the people themselves, it turns out that people do not shop unconditionally in those places that are closest by, but 'which are familiar': According to own observations and interviews with people living and shopping in the border area (winter 2001/2002; summer 2002; i.e. after the introduction of the Euro as common currency), Germans living in the German part of a small Dutch-German cross-border village (Wyler), for example, go rather shopping daily goods in other German places (4 or 13 kilometres away respectively) than to the Dutch shop around the corner (and 'across the border'). In some cases people even do not 'know' that there is a shop in the Dutch part, or they know (passively), but simply never consider shopping there (actively). On the other hand, Dutch people who moved to one of the enclaves on the German side still go shopping daily goods in the Netherlands – though there is a supermarket across the street (see also Rengers 2002). The latter is a form of 'double' or 'to-and-fro border crossing' which also makes clear that the borderline is not important as such, as a line on the ground, but as 'cognitive barrier' in one's mind that refers, among other things, to feeling at home or abroad. This 'double border crossing' results also in a sort of non-crossing, or non-practice of border crossing.

These findings, however, are quite the reverse concerning 'fun shopping' (e.g. clothes, furniture) or buying non-daily goods with remarkable price differences (e.g. alcohol), for which people actually do cross the border (see also Van der Velde 2000b). But in general, similar results were found with respect to leisure activities: sports on a rather daily basis (e.g. in sports clubs or fitness-centres) take place 'at home', whereas exceptional events such as ambitious (one-day) cycling tours gain significance for the sportsmen if they include the excitement of border-crossings. Finally, these differences in everyday and exceptional practices considering the border (-crossing) can also be found in the spheres of entertainment and education ('cultural life'). As I have outlined elsewhere, there is hardly any cross-border exchange of programmes and visitors with respect to 'ordinary/popular' culture (e.g. cinemas) (Strüver, forthcoming). Conversely, it is common to cross the border for special events such as pop-concerts or open-air festivals. Against this background I would argue that the border is crossed for rather uncommon tours that are 'planned', but not for *everyday* routines and practices.

The borders introduced here as 'barriers in people's minds' are 'cognitive-imaginative borders' insofar as they are (re)produced by and (re)produce narratives and images as well as imaginative and narrative *spatialities*, many of them related to stereotypes and prejudices, creating feelings of strangeness when being in the neighbouring country. This kind of everyday practice of not crossing the border, of cross-border *im*mobility, or even 'performed immobility', is not a general one; this immobility is rather related to the border, than to mobility as such. Against this background, I now turn to images and imaginations related to the border on the one hand and to the significance of narrativity for everyday life on the other.

### Six, Five: 'Ground control' – representations and imaginations

Apparently, spatial borders in general and the ones between the EU member states in particular are not on the 'ground' any longer. They are increasingly accepted as socially constituted and discursively produced phenomena, as real-*and*-imagined affairs, as constructed by representations and as reconstructed by people's practices and imaginations.

Relying on theories of representations I refer to the assumptions that boundaries and regions are not naturally given, self-evident entities, but rather representations (see, for example, Jones and Natter 1999; Paasi 1996; Sidaway 2002). Both regions and boundaries are socially constructed processes with contested meanings and the construction of meanings is embedded in social relations. Representations, in their turn, are products of their social context; that is to say that the discursive and materiality of boundaries are not oppositions but mutually constitutive. Their materiality is produced and interpreted by representations; they are reality through representations. To put it another way, it is only through the *linkage* of the discursive construction of boundaries (as representations) and their material effects (that reproduce the construction) that they become meaningful. Their materiality is attached to representations and the former both embeds and conveys meaning(s) only *through* representations. Thus, what is regarded as a 'material' boundary is the outcome of and interpreted through representations.

Representations such as narratives and images are, in turn, materialisations of social relations embedded in space. For social spaces and boundaries are constructed and *practiced* through texts and images, it is only through representations that they are perceived, constituted and experienced – though usually 'representations disguise the practice that organise them' (de Certeau 1986:203). In short, there is another mutual relation, that is the one between representations and practices, i.e. practices produce representations, but representations are also part of practices through which realities are established. Representations therefore perform realities and they are significant *in* practice and *as* practice – since they shape people's everyday-life. Their meanings are explicit or implicit, conscious or unconscious, common sense or 'nonsense', and they might be carried in everyday speech, popular culture and high art, transmitted by TV, music, internet etc.

By representations, however, I obviously do not refer to mimicry and reflections, or merely carriers of meaning, but to meanings that occur and are constructed through representations that arise from various practices and interests. Since representations cannot be reduced to representational *forms*, the ways they are read, seen and interpreted by audiences are equally important. A representation as such does not reveal much about its ways of reception and perception by its (various) 'users'. These 'users' *produce* meanings through their (signifying) practices.

This notion of 'users' as 'producers' goes back to Michel de Certeau who was particularly concerned about the significance of narrativity for everyday life. In a chapter titled 'The Establishment of the Real' (1984) he postulates that narrations are about what-is-going-on, that they constitute our thoughts and that this narrated reality constantly tells us what must be believed and what must be done. People learn about the world and themselves through narratives. Moreover, narratives have the strange but strong power to transform seeing into believing, of fabricating realities out of appearances. They are acts of making, i.e. they do not tell something, but *create* something. What is more, de Certeau put an emphasis on the interrelations of everyday practices and narrations *with space*. According to him, narratives are signifying practices that *invent and organise spaces* and thus organise and regulate movements in space. As a result, narratives are always about spatial practices on the one hand and space is part of everyday practices on the other. Narratives produce 'geographies of action' (1984:116), and practices are 'modes of operating or schemata of action' in which 'users' 'reclaim' the space organised by socio-cultural production (ibid.: xiv–xv). In a similar way, narratives play an everyday, but decisive role for marking out boundaries (including the justification to cross them) and on the whole, 'there is no spatiality that is not organized by the determination of frontiers' (ibid:123). De Certeau even refers to boundaries as stories that 'traverse and organise places, that select and link them together' (ibid.:115). This is what he calls the operations of marking out boundaries; operations that consist of stories that are derived from fragments drawn from earlier stories. And stories are more than a fixation of earlier versions in the past, they are creative acts and performative forces; they are polyvalent because of their mix of many different stories.

However, what can we learn about borders in people's minds from borders as representations? As, for example, narrated in novels, performed in theatre plays or presented in exhibitions, cartoons and on TV? Narratives and images do not only depict or reflect the border. Since they are interpreted differently, they play crucial roles in creating it, in giving them meanings and establishing realities.

This kind of narrated, but established reality can also be found in Benedict Anderson's conception of *Imagined Communities* (1983) in which he introduced the idea that the space and time of the modern nation are embodied in the narrative culture of the realist novel. And Homi Bhabha (1990) also addresses the *Nation as Narration* and draws not only attention to the narrative but also attempts to alter the concept of the narrative itself: Because we find narration at the centre of the nation, we need to include the performativity of language in the narratives of the nation and the insights of post-structuralist theories of narrative knowledge (textuality, discourse, enunciation etc.). However, what both Anderson and Bhabha focus on in their explanations is the acknowledgement that the differences between nations rely on the principle of 'us' versus 'them'. Referring to (post)structuralist's differential concepts such as Derrida's *différance* (1991), it becomes clear that meanings are established in and derived from the dominant use of binary oppositions in narratives. These oppositions are a fundamental operation to the production of meaning, that is to say that meanings are defined by their relation to what they are not.

I argue that this principle can be transferred easily to the nations' edges, their boundaries. The urge to emphasise a difference, to state 'I am because I am *not*' refers to the well-known process of identification – which is always a process of distinction, of marking and making boundaries. Consequently, the mechanism of différance is even more effective *at* the borders, not within nation states. Related back to the Dutch-German borderland, to representations mirroring and constructing Dutch-German relations on the one hand and stereotyping as representational practice and boundary maintenance on the other, meanings are differentially defined along this border, mostly by what one/the other side is not. Renckstorf and Lange (1990), for example, have emphasised in their research results on Dutch-German representations that this urge to differentiate, to stress where, what and who one is and *is not*, gets stronger the closer people live at the border.

Mutual narratives and images about Germany and the Netherlands – as they are 'mediated' in novels, newspapers, cartoons and on TV, for example, (still) very much rely on the past that is related to World War II. During the five years of German occupation, many Dutch cities were devastated and the Nazis controlled and destroyed people's lives. Whereas political conditions and economic co-operation have returned to the normal in the meantime, the socio-psychological relations remain difficult. Depending on the occasion, they can be characterised within the range of 'sensible scepticism', 'still very prejudiced', 'kind indifference' and 'friendly ignorance'. Summarised superficially, but yet dominating popular culture in general and TV-, radio- and print-commercials as well as schoolbook-, newspaper- and magazine-cartoons in particular, Germany as a country is too big, too close and too unreliable for the Dutch and the Germans are fat, xenophobic and have no sense of humour. Vice versa, the Netherlands is reduced to beach holidays, cheese, greenhouse vegetables, tulip fields, windmills, caravans and drugs (Keim 1998; Linthout 2000; Piel 2000).

Peter Groenewold (1997), who has been working on Dutch-German relations for more than twenty years, states that Dutch wartime experiences resulted in an active ignorance of Germany, sometimes so strong that Germany does not exist in Dutch perception. On the other hand, the Netherlands is only one out of nine nation-states bordering on Germany (and on top of that a rather small neighbour) so that the other way round results in a passive, but yet ignorance. Groenewold's arguments are based on an analysis of innumerable narratives of Dutch-German 'encounters' spanning the past 300 years. It has resulted in an interpretation of Dutch-German relations that could be characterised by their close neighbourhood, the dualisms of closeness and distance, friendship and enmity, big and small country. Against this background, Groenewold has developed a 'Dutch-German interpretation array' that consists of differentially defined characteristics. These are (among many others) 'big/ small country, top-down/bottom-up politics, warlike/peaceable mentality, conflict/ consensus attitude, individual/collective emotions, enjoyment/asceticism as way of life, idealism/realism as political philosophy, guilt/morals as predominant feelings' and can be condensed to 'moral geopolitics of guilt, shame and pride between Germany and the Netherlands' (Groenewold 1997:180–83). But beyond the trauma of wartime occupation the relations also remain ambivalent – for the (assumed) tolerant, but moralistic Dutch active ignorance and the expected, but unspoken German official apology and consequently passive ignorance seriously interfere with any attempt at rapprochement. The present occasions in popular culture and everyday life were all these 'funny stereotyped images' and 'serious categorisations' mentioned above 'crash and clash' are, for instance, the sensitive football matches between the Dutch and German national teams (Verheyen 1993).

A 'confrontation' with that kind of images (and accompanying narratives) does not result in a simply takeover of stereotypes and prejudices, or in concrete 'formations' of people's opinions and attitudes. Yet, being confronted with them on a day-to-day basis influences people at least in the sense that they have to deal with them – often unconsciously. But still, people do not watch/listen passively, they rather 'make' something out of it and these (re-)makings are forms of signification that constitute familiarity and the respective Dutch/German 'common culture'. Representations thus perform and organise commonalities and realities and people rely on familiar representations in order to articulate their identities (de Certeau 1997). A very well-known contemporary narrative about 'the Germans' that can be found in various cartoons illustrates a (German) couple or family, demarcating 'their' piece of (Dutch) beach by building a sand-wall and setting up German colours. This image conjoins several aspects of Dutch-German relations and stereotypes associated with them: The Germans 'occupy' the Netherlands (again), they demarcate their 'territory', they keep people out, they simply hang out, drink beer, get fat and do not care about anybody except for themselves … All these 'attributes' go back to World War II, but refer also to more recent events such as racist attacks in Germany during the 1990s and the (in)capacity of the 'German society' to deal with them, to 'the Germans' habit to 'overlook' those incidents and continue to live their normal lives (which is mainly, of course, drinking beer). Concentrating on dominant narratives and images of 'the Dutch', most of them deal with agricultural products (tasteless tomatoes, typical tulips, cheese etc.), Dutch tolerance (towards drugs, abortion etc.) and eccentricity (e.g. caravan- and bicycle-mania) – all of them referring to Dutch smallness and therefore cuteness in one way or another and implying that 'the Dutch' do not need to be taken too seriously … Everybody knows that these 'attributes' are 'not true', not to be generalised etc. However, there is an underlying pattern that usually represents 'the Germans' as imperious and authoritarian and 'the Dutch' as frivolous and grotesque (Keim 1998). As a result, these stereotyped 'characteristics' are real in a way that they are everywhere – they circulate in representations and in people's minds, they are played with, varied and re-employed, they get rejected and yet applied to any kind of present events and encounters.

Especially Groenewold's findings about the Dutch-German relations characterised by the dualism of 'proximity and distance' is an issue that kept Georg Simmel busy in his works on space and social forms. He asserted that the production of space is at the same time the construction of social distinctions and that this is especially true for spatial boundaries. Simmel was maybe the first one who emphasised that a territorial border is not a spatial fact with social impact but rather a social fact in spatial shape (Simmel 1992).

The (general) urge people feel to belong, to create their 'own space', to separate, to differentiate and to demarcate, and their attempts to put this into practice was another topic of Simmel. He argued that being and feeling socially close does not require spatial proximity. On the other hand, people who are spatially close to each other, but belong to another group, are often socially remote. This phenomenon mirrors the almost always-existing synthesis of nearness and remoteness, of closeness and distance, and is described by Simmel as social interaction that is lived as involved difference. Being spatially close, but socially remote is being neither insider, nor outsider, but 'near and far at the same time' and 'a strange element within a group' (Simmel 1950).

And finally, in a short essay, he particularly elaborated on borders and outlined an understanding of both bridges and doors as images of boundaries that both separate and connect (Simmel 1997). Whereas the bridge symbolises the connection between what is separated, the door as metaphor is even more significant in illustrating that connection and separation are the two sides of the same act. Consequently, the door is also an image of the border, as the blocking *or* permitting effect of a border.

With respect to the removed border between Germany and the Netherlands this makes clear that the people are spatially close, but socially remote because they belong (or: feel they belong) to another group ('the Dutch'/'the Germans') and might experience feelings of strangeness when being on the 'other side'. But is always *people* who limit, separate and border, who create 'social distance' – without being limited by a border themselves. Though people could build bridges, step through the door and cross the border easily, they have a hard time to reach out and get in touch with each other. The doors within the EU are wide open and EU-citizens are not only permitted, but invited to cross the internal borders. Yet, the thresholds seem to be too high – and result in what I have introduced as non-practice of border crossing earlier. This notion of cross-border immobility mentioned repeatedly makes me turn to theoretical ideas of this 'performed immobility' in relation to everyday practices, i.e., to practices as representations and practices produced by the 'use' of representations.

**Four, Three: 'The spaceship knows which way to go' – embodied practice and performances**

Fantasies about borderless mobility in general, but particularly within the EU, draw upon ideas and assumed ideals of 'boundlessness'. The Dutch-German border's 'condition' is one of a postulated borderlessness, a notion that aligns it with the imagination of a 'borderless world' or a 'deterritorialised space of flows' associated with a somewhat empty space filled with free-floating subjects. This is, of course, a matter of interpretation: The strange, but nevertheless established fact is that the promoted borderlessness on the one hand and cross-border co-operation on the other are inconsistent with each other – for, as I have argued elsewhere, co-operation *across a border* relates to a still existing and even a rather 'closed' border (Van Houtum and Strüver 2002). Notwithstanding this objection and in order to 'ground the (assumed) flows' across the (removed) Dutch-German border, I will now move from representations *of* the border and the border *as* representation to (everyday) practices. This attempt to link representations with practices, or to 'ground' narratives and images in practices presents the aim of approaching the 'odd spaces of immobility' from a new angle.

Neither images nor imaginations as such, nor their producers and consumers, just flit around in a vacuum. On the other hand, 'practices' (whether or not influenced by representations) should not be reduced to the 'ground' of action, as just happening (voluntarily or rationally intentional) or as outcome of broader social structures. Against this background, and sympathetic as I am to the various linguistic, semiotic, cultural, representational, pictural and spatial turns in social theory (accepting and working from their presuppositions and effects), I am also concerned about 'embodied' practices. For the remainder of this contribution, I try to 'locate' the everyday non-practice of border crossing (performed immobility) without abandoning the representational perspective. In addition, I adopt a way of thinking that 'puts practices into practice and lets them act', and in so doing I join the bandwagon of embodiment and performativity in order to 'ground the flows', to put life into representations.

This is, of course, a schema Nigel Thrift already proposed some time ago (1996; 1997). In what he terms *non-representational theory* (NRT), a rich combination of Bourdieu, de Certeau, Foucault, Deleuze and Guattari, actor-network-theory and many other approaches, Thrift outlines a theory of practices that links performativity and embodied practices. Thrift's goal is to build a theory of everyday practices that shapes the attitude of human beings. Thus it is not 'a project concerned with representations and meaning, but with the performative "presentations", "showings" and "manifestations" of everyday life' (Thrift 1997:127).

Within this framework, Thrift has examined social theories, which conceptualise society neither as an underlying code nor as an inscribed surface, but rather as a more or less spatially and temporarily extensive network of practices. And what is more, he is deeply sceptical about representations, for he understands *representations in opposition to lived experiences and materiality*: 'A hardly problematised sphere of representation is allowed to take precedence over lived experience and materiality, usually as a series of images or texts which a theorist contemplatively deconstructs, thus implicitly degrading practices' (Thrift 1996:4). Non-representational theory thus abandons representations to a certain degree, for it is assumed that they are inadequate to recall ordinary lives. And this label of inadequacy is founded on the idea that representations such as written and spoken texts and visual images exclude 'embodied practices'.

Non-representational ways of thinking, on the other hand, express the fact that practices constitute people's sense of the real and emphasise the practice of everyday life as *embodied*. NRT is concerned about thinking with the entire body, all its senses – not only the visual. That is why the theory is sceptical about the linguistic turn, suggesting that it too often cuts people off from human practices. Since Thrift places emphasis on the body, he also suggests use of the term 'embodiment' – as embodiment is a process, is practical. Thus, embodiment is perceived in terms of involvement, it is being in relation to a world (Thrift 1996; 1997).

Catherine Nash (2000) and others have argued that ideas of performance, embodiment and practice are valuable in the sense that they challenge textual analysis (interpretations that focus on the representations of meaning in images and texts). But for Nash, NRT's rejection of texts and focus on bodily-practices can also be read as an attempt to re-build old Cartesian dualisms between mind and body, thought and practice, which many scholars – including Judith Butler and various other feminist and/or representational theorists – have claimed to overcome (see, e.g., Butler 1990; 1993; 1997; Grosz 1995; Haraway 1991). And what is more, 'nonrepresentational theory does not seem to allow room for considering visual and textual forms of representations as practices themselves' (Nash 2000:662). Nash, therefore, concentrates on the advantages that could be gained in turning from representations to performances and practices, from the 'new theoretical vocabulary of performance on the one hand and the imaginative and material geographies of cultural performativity and embodiment on the other' (ibid.:654). Not surprisingly, she refers to Butler's theory of performativity (Butler 1993; 1997) in order to overcome the distinction between mind and body.

Despite having a different perspective and focus than Thrift, Judith Butler's theory of performativity also constantly stresses the human body, embodied subjects and practices. But Butler conjoins representations and practices *together*, rather than keep them apart in the manner that Thrift does. I therefore accuse Thrift of having a much too limited/restricted understanding of representations (reduced to texts and images in their narrowest sense). As I have mentioned earlier, I rather imagine *representations as arising from and being produced through practices*. In what follows, I therefore bring practices and representations together through the notion of performativity – at first on a theoretical level and then 'applied' to the Dutch-German borderland.

Although Butler's key concerns initially revolved around issues of gender and sexual identity, the idea of performativity has been applied also to the construction of national, ethnic and racial identities (e.g. Butler 1997, Fortier 1999), and I propose that it could be transferred to any process of enacted identification. Butler has developed a linguistic definition of performativity ('speech act'), one that explicitly rejects theatrical notions of performance and is also opposed to a merely psychoanalytical account of performance. In her understanding, performance must always be linked to performativity; whereas performance is what individuals do, say, or act out, performativity is the constant repetition of norms in citational practices, which reproduce and/or subvert discourse and become known through representations. For Butler, performativity is 'the reiterative and citational practice through which discourse produces the effects that it names' (1993:2). Performativity is thus constitutive of representations and their meanings as well as of embodied identities and practices (and, vice versa, a performance's reproduction or transformation is constituted by representations), and so, we should not think of practices, performances, texts and images separately, but as a whole.

According to Butler (1997), an image, a text or a single word do not only signify a thing, but these forms of representations as significations are also the enactment of the thing and that is why a performative act is both signifying and enacting. A performative act does not succeed because of an underlying intention that governs the speech act, but because the act is an echo of previous acts or actions 'and accumulates the force of authority through the repetition or citation of a prior and authoritative set of practices' (1997:51). This idea implies that the speech act does not only take place within a practice, but that the act itself is a ritualised practice and therefore the performative draws on and covers the constitutive conditions that mobilise it. In short: (1) thinking signification as performance includes its enactment, and (2) the effects of repeated citations are materialisations.

More recently, and following her earlier rejection of natural, pre-given and fixed gender and sexual identities (1990), Butler (1997) has elaborated (implicitly) on the criticised mind-body dualism by reworking Pierre Bourdieu's considerations regarding the mode of generation of practices. In this, she has explicitly rejected Bourdieu's distinction between the linguistic and the social (a distinction that also can be found in Thrift's NRT), as she doubts that the body (as the place where performative commands are received, inscribed and carried out at the same time) and practices can be approached by separating social and linguistic dimensions. It is thus obvious that Butler's conception of the performative act is not reduced to official language and discourse. It is rather a 'social ritual, a silent and insinuating practice' that is not acted out by a pre-given subject and that results in effects beyond the control of a speaking subject. Performativity is about the 'reiteration of norms which precede, constrain and exceed the performer' (1993:234). Understanding identities as performative implies that they are constructed by the 'very expressions that are said to be their results' (1990:45). The performative act is one of the ways the subject is called into being, a part of subject formation, of its ongoing political contestations and reformulations (in the sense of anti-ritualistic transformations and change). Yet, Butler conceptualises performativity not as mere routine, as replicas and reiteration of practices, but citationality 'through the invocation of convention' (1993:225–6).

### Two, One: 'It is time to leave the capsule' – Representational space and 'grounded practice'

In order to locate and ground representations in practices, to *link* them with practices and to relate them back to an 'imagined' borderless European space of mobility on the one hand, and to a 'real' non-practice of border-crossing between Germany and the Netherlands on the other, I argue that imaginations about 'spaces of flows' and 'borderlessness' must be 'grounded'. In order to approach a (provisional) conclusion, I will turn to conceptual thoughts about 'Space Oddity' for a while and subsequently re-connect those theoretical ideas about performativity, everyday practices and representations with the 'performed immobility' along the Dutch-German border.

Performativity as the reiteration of practices, as citation of conventions, operates through embodied practices as bodily movements in space that recall and reconnect with places. If I (over)emphasise the notion of place in this context, the non-practice of border crossing turns out to be a conventional practice. Moreover and extending Butler, a performative act in this context is a 'social ritual, a silent and insinuating practice'. This could easily be linked to Michel de Certeau's idea of everyday practice as consumption (1984). In this sense, the consumers of representations are (active) users and producers. Despite the clandestine character of this kind of production, they are *practitioners* – whose practices are embodied through bodily movements.

Both (embodied) practices within space and the production/construction of space ('as such') *'take place'* through representations, through performative acts, through acts of narration, visualisation, imagination. Yet, those performative practices cannot be reduced to representations as illustrations. Rather, they generate a representational space.

To wind up this risky thought experiment, bringing my odyssey about space oddity to a close, I will stop to 'float in a most peculiar way', 'leave the capsule' and call for 'ground control' (Bowie 1990). But, since I am not 'Major Tom', ground control is also someone different: Henri Lefèbvre will be the friend I need for my last imagination (for the time being, of course). This 'representational space', then, is part of Lefèbvre's conceptual triad (1991b), a way of understanding the process of producing space and a way of thinking practices and representations *together*. He calls 'spatial practices' people's perceptions that influence their repeated actions and movements in space that make up the day to day. Spatial practices are made up of daily routines, including spatial routines such as networks of routes between work, shopping, leisure activities, home etc. These spatial routines do not only take place in space, but also produce (a society's) space. The ability to produce space, in turn, includes its modification and change (Lefèbvre 1991a, 1991b). 'Representations of space', then, are discursively produced by officials through signs and codes, and 'representational spaces' refer to the use and experience of space. Representational spaces – or 'spaces of representations' – are the spaces of everyday life that are experienced, lived and produced by their 'users' through representations, which are 'linked to the clandestine or underground side of social life' (Lefèbvre 1991b:33). And although Lefèbvre was more concerned about *urban* spaces and about 'change' in a rather empowering sense, i.e. about counter movements that invent new (urban) spaces as sites of resistance, I will proceed transferring this idea to the Dutch-German borderland in the concluding section.

Summing up, one could say that the use and consumption of space take place through representations. But spatial practices and their products are not determined by those representations. Practices, understood as activities of *'making do'* as well as performative acts, are both embedded within ongoing social processes – including transformation and change. It is only because of the interplay with practices that representations become alive. In short, performative acts as embodied practices and bodily movements are ways of consuming representations and producing spatial practices, as well as producing and changing 'representational spaces'.

## Liftoff: 'Stepping through the door' – conclusions

In this chapter I have outlined that the official European discourses on mobility are less influencing on people's actual practices with respect to crossing borders than representations. But although representations such as narratives and images are rather clandestine than obvious in their character and disguise the ways they come into effect, they are highly effective for spaces of representations as the Dutch-German borderland. They effect and affect the production of spatial practices, mobility as well as immobility.

'Spaces of representations' along the Dutch-German border are those spaces that are made of people's spatial patterns, emerging from their everyday practices and routines, their spatial movements between work, shopping, sports etc. Yet, both practices and spaces are also produced by representations that are mediated mainly by popular media (TV, newspapers) and interpreted by way of consuming them. Representations of Dutch-German relations and their interpretations invent (social) spaces and their borders; they regulate people's movements in space. Narratives and images thus make the border, which marks and delimits people's spheres of daily practices.

I have argued that 'performed immobility' is the non-practice of border crossing. This refers to the maintenance of the border, to everyday practices produced by representations along the Dutch-German border that (re)produce imaginative spatialities about the other side of the border and the people 'over there'. The cross-border regions between Germany and the Netherlands are anything but 'regions without borders' – despite the European Union's efforts to propose and establish borderlessness. Moreover, to accept that belonging has an affective dimension and that the performativity of belonging 'cites' the norms that reproduce a group (or nation), requires to be sensitive for 'imagined communities' and their cultivated sense of belonging – dependent on the ways representations such as narratives and images as well as imaginations are 'used' and result in practices. All this is linked to people's and communities' attachment to and demarcation of places. The hype to feel part of a group and to demonstrate a sense of belonging (e.g. with varying forms of representations) includes the hype to differentiate, to create 'codes' and use norms to present and establish differences, to mark and maintain borders – as the 'double border crossings' of people living in enclaves show.

Following the idea of 'moral geopolitics of guilt, shame and pride between Germany and the Netherlands' that are present during football matches, but also when driving on roundabouts and motorways with 'the other' number plate (to mention but one example), the border still exists. It is a real-and-imagined border, and thus a cognitive-imaginative border that obstructs actual cross-border mobility. Among other things, its affective dimensions refer to the 'conventions' of Dutch-German dislike – and those 'conventions' are related to being spatially close, but socially remote.

The perspective introduced in this contribution offers new insights *why* people do not cross the border as the most natural thing in the world and thus do not correspond with the European Union's ideas about cross-border or even borderless interaction and (dominant) conception of people's practices as rational actions. The challenge 'Stepping through the door' therefore includes letting the measuring of economic transactions and flows of people go when dealing with European 'integration' – or at least keeping it on the sidelines and being concerned about societies as such and people themselves instead. For, the relaxation or even removal of border controls and the progressive establishment of cross-border infrastructure seem to be less important and influencing than those clandestine, but popular narratives and images.

However, the simple invitation to 'open doors' and to step through them – read: to cross the border – is obviously somewhat naïve. Apart from the fact that this approach of 'open doors' is not meant to be applicable to any border, but rather limited to those that are said to be removed, to those insignificant or even 'boring borders' for EU-citizens within the EU, it is not far-reaching enough. Yet, I have no concrete prospects for lowering the thresholds that the border's affective and cognitive meanings create. For a start, I would suggest to apply the notion of the performative also to the border itself: It does not exist and has no meanings outside its processes of production and reiterations. It is maintained as – and thus remains – an imagined border; an imagined border that has the status of the real.

## Note

1    David Bowie (1990) 'Space Oddity'. On CD 'Space Oddity', EMI (EMI Music Germany).

## References

Anderson, Benedict (1983): *Imagined Communities. Reflections on the Origin and Spread of Nationalism*. London: Verso.

Bhabha, Homi K. (1990): *Nation and Narration*. New York: Routledge.

du Bois-Reymond, Manuela (1998): 'European identity in the young and Dutch students' images of Germany and the Germans'. Comparative Education 34(1), 27–40.

Busse, Gerd and Jos Frietman (1998): 'Grenzüberschreitende Arbeitsmobilität in der Euregio Rhein-Waal und in der euregio rhein-maas-nord'. In: de Gijsel, Peter and Hans-Joachim Wenzel (eds): *Mobilität und Kooperation auf grenzüberschreitenden Arbeitsmärkten: Deutschland – Niederlande*. Osnabrück: IMIS Beiträge (No. 9), 37–61.

Butler, Judith (1990): *Gender Trouble. Feminism and the Subversion of Identity*. London: Routledge.

Butler, Judith (1993): *Bodies that Matter. On the discursive Limits of 'Sex'*. London: Routledge.

Butler, Judith (1997): *Excitable Speech. A Politics of the Performative*. London: Routledge.

de Certeau, Michel (1984): *The Practice of Everyday Life*. Berkeley: University of California Press.

de Certeau, Michel (1986): *Heterologies: Discourse on the Other*. Minneapolis: Minnesota UP.

de Certeau, Michel (1997): *Culture in the Plural*. Minneapolis: Minnesota UP.

Derrida, Jacques (1991): 'La Différance'. In: Kamuf, Peggy (ed.): *A Derrida Reader*. Columbia: UP, 59–79.

Van Dijk, Jouke and Teun-Jan Zanen (2000): 'Grensoverschrijdende samenwerking en de arbeidsmarkt in de Eems-Dollard regio'. In: Boekema, Frans (ed.): *Grensregio's en arbeidsmarkten*. Assen: Van Gorcum, 59–82.

Dortmans, Koen (2002): 'Een verenigd Europa begint bij jezelf'. In: *LUX maandprogramma* maart 2002. Nijmegen.

Fortier, Anne-Marie (1999): 'Re-membering places and the performance of belonging(s)'. *Theory, Culture and Society* 16(2), 41–62.

Groenewold, Peter (1997): *'Land in Sicht': Landeskunde als Dialog der Identitäten am Beispiel des deutsch-niederländischen Begegnungsdiskurses*. Groningen (Proefschrift Rijksuniversiteit Groningen).

Grosz, Elizabeth A. (1995): *Space, Time and Perversion: Essays on the Politics of Bodies*. London: Routledge.

Haraway, Donna (1991): *Simians, Cyborgs and Women: the Reinvention of Nature*. London: Free Association Books.

Houtum, Henk van and Strüver, Anke (2002): 'Borders, Strangers, Doors and Bridges'. *Space and Polity* 6 (forthcoming).

Janssen, Manfred (1999): 'Obstacles and willingness for cross-border mobility'. In: de Gijsel, Peter; Janssen, Manfred; Wenzel, Hans-Joachim and Michael Woltering (eds): *Understanding European Cross-Border Labour Markets*. Marburg: Metropolis, 143–164.

Janssen, Manfred (2000):' Borders and labour-market integration'. In: van der Velde, Martin andHenk van Houtum (eds): *Borders, Regions, and People*. London: Pion, 47–68.

Janssen, Jacques; Spille, Hilde and Cor Baerveldt (1996): 'Wetten, praktische bezwaren en weemoedigheid. Een onderzoek naar mobiliteitsverhinderende factoren onder Duitse en Nederlandse jongeren'. In: Renckstorf, Karsten and Nol Bergmans (eds): *Nederlanders en Duitsers*. Nijmegen: Stichting Centrum voor Duitsland Studies, 43–56.

Jones III, John Paul and Wolfgang Natter (1999): 'Space "and" Representation'. In: Buttimer, Anne; Brunn, Stanley D. and Ute Wardenga (eds): *Text and Image. Social Construction of Regional Knowledges*. Beiträge zur Regionalen Geographie 49, 239–247.

Keim, Walther (and Botschaft der Bundesrepublik Deutschland in Den Haag) (1998): *Hallo Nachbar! Dag Buurvrouw! Deutsch-niederländische Beziehungen in der Karikatur*. Osnabrück: Secolo.

Kramsch, Olivier (2002): 'Re-Imagining the Scalar Topologies of Cross-Border Governance: Eu(ro)regions in the Postcolonial Present'. *Space and Polity* 6 (forthcoming).

Lademacher, Horst (2000): 'Reizende Nachbarn – Die Beziehungen zwischen Niederländern und Deutschen'. In: Haus der Geschichte der Bundesrepublik Deutschland (Hg.): *Deutschland – Niederlande. Heiter bis wolkig*. Bonn: Bouvier, 60–73.

Lefèbvre, Henri (1991a): *Critique of Everyday Life*. Volume I. London: Verso.

Lefèbvre, Henri (1991b): *The Production of Space*. Oxford: Blackwell.

Linthout, Dik (2000): *Onbekende Buren*. Amsterdam: Atlas.

Nash, Catherine (2000): 'Performativity in practice: some recent work in cultural geography'. *Progress in Human Geography* 24(4), 653–664.

Paasi, Anssi (1996): *Territories, Boundaries and Consciousness*. Chichester: John Wiley.

Paasi, Anssi (2001): 'Europe as a process and social discourse: considerations of place, boundaries and identity'. *European Urban and Regional Studies* 8(1), 7–28.

Piel, Alexandra (2000): *Skurril, boomend, traditionsbewusst und sparsam. Das Nachbarland Niederlande im Spiegel der deutschen Presse*. Düsseldorf: NED.WORK.

Renckstorf, Karsten and Olaf Lange (1990): *Niederländer über Deutsche. Eine empirische Studie zur Exploration des Bildes der Niederländer von Deutschen*. Nijmegen: Stichting Centrum voor Duitsland Studies.

Rengers, Merijn (2002): 'Duitse villa's prooi Nederlandse koopjesjager'. In: *De Volkskrant* 03.08.02, p. 1.

Scott, James W. (2000): 'Euroregions, governance, and transborder co-operation within the EU'. In: Van der Velde, Martin and Henk van Houtum (eds): *Borders, Regions, and People.* London: Pion, 104–115.

Scott, James W. (2002): 'A Networked Space of Meaning? Spatial Politics as Geostrategies of European Integration'. *Space and Polity* 6 (forthcoming).

Sidaway, James D. (2002): 'Signifying boundaries: detours around the Portuguese-Spanish (Algarve/Alentejo-Andalucía) borderlands' [forthcoming in Geopolitics].

Simmel, Georg (1950): 'The Stranger'. In: Wolff, Kurt (Trans.): *The Sociology of George Simmel.* New York: The Free Press, 402–408 [1908].

Simmel, Georg (1992): 'Der Raum und die räumliche Ordnung der Gesellschaft'. In: *Soziologie.* Untersuchungen über die Formen der Vergesellschaftung. Gesamtausgabe, Bd. 11. Frankfurt/Main: Suhrkamp, 687–790 [1908].

Simmel, Georg (1997): 'Bridge and door'. In: Leach, Neil (ed.): *Rethinking Architecture. A Reader in Cultural Theory.* London: Routledge, 66–69 [1909].

Strüver, Anke (forthcoming): 'Presenting Representations – On the analysis of narratives and images along the Dutch-German border'. In: Berg, Eiki and Henk van Houtum (eds): *Mapping Borders between Territories, Discourses and Practices.*

Thrift, Nigel (1996): 'Strange country: meaning, use and style in non-representational theory'. In: *Spatial Formations.* London: Sage, 1–50.

Thrift, Nigel (1997): 'The still point'. In: Pile, Steve and Michael Keith (eds): *Geographies of Resistance.* London: Routledge, 124–151.

Van der Velde, Martin (1999): 'Searching for jobs in a border area: The influence of borders in a Dutch Euroregion'. In: de Gijsel, Peter; Janssen, Manfred; Wenzel, Hans-Joachim and Michael Woltering (eds): *Understanding European Cross-Border Labour Markets.* Marburg: Metropolis, 165–181.

Van der Velde, Martin (2000a): 'On the value of a transatlantic dialogue on border research'. *Journal of Borderland Studies* XV(1), 281–290.

Van der Velde, Martin (2000b): 'Shopping, Space, and Borders'. In: Van der Velde, Martin and Henk van Houtum (eds): *Borders, Regions, and People.* London: Pion, 166–181.

Verheyen, Dirk (1993): 'The Dutch and the Germans: beyond traumas and trade'. In: Verheyen, Dirk and Christian Soe (eds): *The Germans and their Neighbors.* Oxford: Westview Press, 59–81.

# PART II
# TERRITORIALITY, MOBILITY AND IDENTITY POLITICS

Chapter 5

# Framing Mobility and Identity: Constructing Transnational Spatial Policy Discourses

Ole B. Jensen and Tim Richardson

## Introduction

This chapter focuses on how spatialities get constructed in spatial policy making, and considers how these construction processes might be conceptualised and analysed. Our overall argument is that the analysis of spatial policy discourses will benefit from using a combined framework of concepts and techniques of discourse analysis, together with an understanding of the social construction of space and spatiality. Drawing from research into the field of European spatial policy we illustrate how new spatial policy discourses create new systems of meaning about space, based in this case on the language and ideas of polycentric urban systems and hyper mobility. We then show how this discourse becomes institutionalised in a new set of spatial practices shaping European space and identity.

We begin by reflecting briefly on the dream of a 'frictionless society', which we use to set up the contemporary philosophical and societal context of a cultural sociology of space. Drawing on debates in critical geography and sociology, we argue that a practice- and culture-oriented understanding of the spatiality and mobility of social life is vital to understanding spatial policy making, in line with the shift towards a 'new spatially conscious sociology' (Sayer 2000:133). The approach hinges on the dialectical relations between material practices and the symbolic meanings that social agents attach to their spatial environment. Socio-spatial relations are conceptualised in terms of their practical 'workings' and their symbolic 'meanings', played out at spatial scales from the body to the global, and in the construction of spatialities of mobility. We introduce three elements of a conceptual framework for understanding such a cultural sociology of space: spatial practices, symbolic meanings, and the politics of scale. The approach, emphasising spatiality and mobility as inescapable components of social life, is intended to sharpen understanding of the specificity of spatial policy making.

We then turn to the more practical task of using these perspectives in the analysis of spatial politics and planning. We do this using a non-textually oriented approach to discourse analysis, exploring representations of space in spatial policy discourses using a triangular framework focusing on languages, practices and power-rationalities. We illustrate how this framework can be operationalised using examples of the framing of mobility and identity in the emerging field of European spatial policy making. We draw in particular upon our analysis of the European Spatial Development Perspective (ESDP) (see Richardson and Jensen 2000; Jensen and Richardson 2001; 2004). Here we analyse the construction of a transnational spatial policy discourse, exploring its meanings, practices and rationales as expressions of a new politics of scale and a new politics of space and mobility. We tentatively discuss how this new transnational spatial politics contributes to a European identity. We end by reflecting on the challenges and possibilities of framing mobility and identity within a cultural sociology of space.

## The dream of the 'frictionless society' – reflections on modernity and mobility

The contemporary dream of the 'frictionless society' is a particular rationality of space and mobility which, we will later argue, is fundamentally shaping the ideas, language, and practices of the new field of European spatial policy making and, through this, placing mobility as the cornerstone of the attempt to imagine European identity. The history of Mankind can be told as one of ever increasing capacities for transcending spatial limits and thus maximising mobility. However, we do not subscribe to a teleological notion of history. Furthermore we subscribe to the claim that what is a liberating increase in mobility for some, creates limits and constraints for others (Urry 2000). On the urban scale of everyday life this is seen as a mobility polarisation between 'cash poor/time rich' and 'cash rich/time poor' commuters (Graham and Marvin 2001:318). Beckmann (2001:17) even goes as far as stating that 'mobilisation needs immobilisation'. Thus modernity may be seen to be intrinsic to the notion of mobility, or stated otherwise modernity can be seen *as* mobility. However, modernity also implies immobility or 'friction', as we would like to term it. Berman uses the myth of *Faust* as an allegory of such modern perception of mobility as progress (Berman 1983:62–63). Part of this Faustian project of ordering the physical environment in accordance with the Modern rationality pays due to ideas of rationalising space (Lefebvre 1974/1991) and overcoming the spatial inscription of disorder and difference (Sennett 1990).

According to Bauman (1998), mobility has become the most powerful and stratifying factor leading to a global hierarchy of mobility. A hierarchy in which a mobile elite transcends spatial barriers at the same time as a growing underclass remains either immobile as inhabitants of the mega-cities' Shantitowns, or is forced into mobility as international refugees. A new restlessness now constructs the idea of a 'state of rest' or immobility as a sign of social degradation (Bauman 1998). In a sense we are now entering a radicalisation of the Modern quest for speed and fast moving objects, as the twentieth century goes down in history as the 'Great War of Independence of Space' (Bauman 1998:8). We have entered a phase of 'liquid modernity' where the logic of dis-embedding works without re-embedding, and de-territorialisation works without re-territorialisation. Location and embeddedness has been replaced by a new disorientation focused on movement. In liquid modernity, power lies in the capacity to 'travel light', in speed and in fast access (Bauman 2000).

Needless to say the desire to transcend spatial barriers is intrinsic to the emergence and development of the capitalist system (Castells, 1996). This has culminated in the new conglomerates of highly flexible technologies and infrastructures which provide the nearest approximation to the frictionless flows of spatial mobilities seen in human history (Graham and Marvin 2001). The expansion of the rationale behind the 'spaces of flows' (Castells 1996) results in a global network connected by nodes and points of modality shift, brought into being by the late twentieth century synthesis of speed, light and power (Thrift 1996:279).

Critics have also advocated limits and alternatives to society's quest for fictionless mobility. The 'Cittá Slow' ('Slow Cities') movement, which originated in the 'slow food' anti-globalisation movement that resisted the 'McDonaldization' of society (Ritzer 2001:177), targets mobility issues directly. The rationale of this movement is to deliberately induce friction in the urban fabric, by enlarging pedestrian areas, building cycle routes, restricting air and noise pollution in city centres, and introducing traffic calming measures. As Urry (2000:55) points out, slowness can be a highly valued way of moving across an environment.

In pursuit of the desire for the frictionless society, mobility has become one of the most potent forms of contemporary power. Struggles over the problems caused by friction, and by attempts to reduce friction, have become one of the most contested areas of spatial policy making. In this chapter we will examine the treatment of frictionless mobility in emerging spatial policy discourse in the European Union. First, we want to introduce a framework which may be helpful in conceptualising and analysing struggles over space and mobility in spatial policy processes, and consequently in shedding light on the nature and meaning of the new politics of mobility shaping European space and identity.

## Towards a *cultural sociology of space*

The fundamental assumption of a *cultural sociology of space* is that analysis must address the dialectical relations between socio-spatial practices and the symbolic and cultural meanings that social agents attach to their environments (he two spheres are separated analytically, not as an ontological statement). Furthermore, by using the term *cultural* sociology of space we wish to underline that the symbolic/cultural aspects of the socio-spatial relation are given as much priority as the material practices (which is the main focus for 'classic' sociology). That is to say, we need to conceptualise socio-spatial relations in terms of their practical 'working' and their symbolic 'meaning'. This dialectical perspective means that the spatiality of social life is thus simultaneously a field of action and a basis for action (Lefebvre 1974/ 1991:73, 191). The specificity of these dialectical relations, as they pertain to actual contemporary spaces and mobilities are enhanced by considering the politics of scale and mobility.

### Spatial practices

The first (analytical) dimension of the cultural sociology of space, dealing with the coercive and enabling effects of socio-spatial relations on social practices, emphasises not only the material dimensions of human agency but also the significance of power. Harvey congenially stresses that social relations are always spatial and exist within a certain produced framework of spatialities, and that this framework consists of institutions understood as 'produced spaces of a more or less durable sort' (Harvey 1996:122). Such spatialised institutions range from territories of control and surveillance to domains of organisation and administration, creating institutional environments within which symbolised and meaningful spaces are produced. In line with the dialectical framework specific places need to be conceptualised in relational terms.

Within a cultural sociology of space, flows and mobilities are addressed as a key dimension in understanding material practices in society. The new mobility forms transforming the spatiality of social life contribute to uneven geographical development 'producing difference' at various spatial scales (Harvey 2000:75–83). This problem of uneven development in the face of globalisation creates a critical problem in framing policy discourses carrying the idea of balanced development. According to Castells the complex dynamics of globalisation can be understood as a dialectical tension between two forms of 'spatial logic' or forms of rationality. The essence of his conceptualisation is therefore a dialectical tension between the historically rooted local spatial organisation of human experience (*the space of places*) versus the global flow of goods, signs, people and electronic impulses (*the space of flows*) (Castells 1996:412, 423). A critical analysis of the representations of space in the emerging European spatial policy discourses reveals them as attempts to frame spaces in line with a particular teleology of European space, which asserts a new space of flows against a space of places.

Turning to the part played by material practices in the production of spaces, Foucault in particular was very interested in how domains of organisation and administration operated through the power relations embedded in local practices, through the 'apparently humble and mundane mechanisms which appear to make it possible to govern' (Miller and Rose 1993:83). This insight suggests the need for close attention to the power struggles within and around the fine grain practices of policy making. The focus of research may turn (for example) towards how commonly used techniques of analysis construct particular forms of knowledge, providing legitimacy for particular spatial strategies whilst marginalising other ways of understanding policy problems. Policy making practices may mask such conflicts, but inevitably they are marked by them.

In these terms, we might conceptualise the emerging field of a European spatial policy discourse as an attempt to produce a new framework of transnational spatialities, which destabilises the traditional spatialities within European member states. This new orientation creates new territories of control, expressed through the new transnational spatial vision of polycentricity and mobility (ESDP). It necessitates new territories of surveillance, manifested in the need for enhanced spatial analysis focusing on new problems at new spatial scales. These new territories are given life by a variety of more or less formal administrative arrangements, including the creation of new cross border policy networks, planning and other institutions of governance, and financial instruments.

*Symbolic meanings*

The second (analytical) dimension of the cultural sociology of space addresses how meaning is attached to the spatiality of social life. In other words, it deals with the question of how representations, symbols and discourses frame the cultural meaning of socio-spatiality. By means of a process of 'social spatialisation' social agents give meaning to spaces through socio-spatial practices and identification processes (Shields 1991:7, 31). Thus social agents 'appropriate' space in terms of ascribing cultural and symbolic attributes to their environment, whilst their spatial practices are simultaneously enabled or restricted by the very quality of this symbolic spatiality. A discursive representation of space prescribes a domain of 'meaningful' actions and thus ultimately provides a regulatory power mechanism for the conditioning and selection of appropriate and meaningful utterances and actions. Parts of these symbolic meanings can be uncovered in what Lefebvre termed the 'representations of space', that is 'the space of scientists, planners, urbanists, technocratic subdividers and social engineers' (Lefebvre 1974/1991:38–9).

Again, it is possible to see social spatialisation as a major activity in the new field of European spatial policy, as physical spaces are attributed new meanings. In the new European spatial vision cities, ports and airports may be re-represented as key nodes in transnational networks. This process of attributing meaning is contested between actors from the local to the European level, given what is at stake in terms of the perceived causal links between connectedness and economic competitiveness. Regions may be re-represented as core, peripheral, urban or rural in the new European geography. And 'local' roads and railways may be re-represented as segments of international high speed transport corridors.

This Europeanisation of spatial representation opens up new questions of identity. A cultural sociology of space, with its double focus on material practices and symbolic meanings, frames these questions of belonging and identity as a matter of both material and cognitive processes. Social identity is here understood as a process of constructing meaning on the basis of pre-given cultural attributes (Castells 1997:6). But, even though social identities can be originated or 'induced' from dominant institutions (e.g. the EU), they will only become identities insofar as social agents internalise them in a process of individuation. This is important both in the context of the debate over 'Europeanness' but also in the context of 'globalisation', and suggests the need to broaden the debate over the nature of EU spatial policy. So policy discourses of European space can be said to do more than carry with them visions and ideas for the transnational functional co-ordination of flows and activities in space. The ESDP in particular, can be seen as both articulating a functional network of regions and nation states in a competitive global region, and injecting spatial and mobility dimensions into the discourse of political integration in Europe, thus potentially 'spatialising' and 'mobilising' the less tangible notion of European identity.

*The politics of scale*

Both spatial practices and the construction of symbolic meanings are played out at particular scales. Scaling should not be understood as either an ontological statement on the profound nature of spatiality or an assumption of fixed hierarchies of places. What it means is rather to notice the power relations and workings of a 'politics of scale' (Brenner 1998). Such politics of scale makes the cultural sociology of space sensitive to the way social agents identify with spaces and places, and sees these spatialities as either enabling or constraining action. In other words, social agents use more or less fixed notions of a spatial hierarchy of nested places in order to navigate reality. In principle the range of this scaling extends from the body ('the geography closest in') to the global (Smith 1993:99).

In the context of European spatial policy, it amounts to seeing the discursive practices of scaling and 're-scaling' from the nested territories of cities and urban regions to the nation state, the new transnational mega regions, and the European Union. The European spatial policy discourses constructed in and across these multilevel arenas are not only expressions of a new politics of scale. They are framed in the context of globalisation and so explicitly articulate a global-local dialectic, manifested in the interaction of multiple and complex localised practices crossing time and space (Harvey 1996).

*Drawing together the cultural sociology of space framework*

Integrating the three dimensions of material practices, symbolic meanings and politics of scale in the cultural sociology of space has several important consequences. Firstly, as Harvey expresses it, representations of space not only arise from social practices, they also 'work back' as regulations on those forms of practice (Harvey 1996:212), thus creating a complex socio-spatial dialectics. In other words, the new spatial visions contained in the ESDP not only express future 'imaginary landscapes', they also 'work' in terms of being vehicles for new forms of transnational policy making. The development of the Trans European Networks (TENs) and the focus on 'missing links' vividly illustrates this point. The vision of a Europe without constraint on the physical movement of goods and people is a representation of a symbolic space of integration and cohesion. However, in order to achieve such a 'frictionless' space there needs to be a more or less fixed definition of the problem. Furthermore, acting according to a given problem definition emanating from such *spatial policy discourse* creates spatial practices. Such practices respond to the new 'problem' of transcending the nation state and physical boundaries such as mountain ranges and stretches of water. They also express the new politics of scale by re-articulating the territory of Europe as a transnational polycentric space connected by long distance, seamless transport networks.

*Secondly*, this approach to the relation between space and discourse also implies that the concept of power is reflected within the theory. As Beauregard reminds us (Beauregard 1995:60), the City does not present itself but is rather represented by means of power relations expressed in strategies, discourses and institutional settings. European space can also be understood in these terms. Thus we subscribe to the numerous conceptualisations of planning as an expression of a 'will to order' (among others see Flyvbjerg 1998; Sennett 1990). The creation of a new activity of European spatial planning can be understood as an expression of a will to order European space, with its emphasis on ideas such as cohesion and balance that articulate a harmonised Europe.

*Thirdly*, and as a direct consequence of seeing planning as the 'will to order', the concept of knowledge and the social relations governing the various claims to valid knowledge are central to the analysis. As Perry reminds us, we should rather think of planning as a spatial and strategic discourse than as a science or knowledge of space (Perry 1995:237). Thus the question is what epistemologies govern the 'knowledge policy' in operation. This leads to a view of planning as more than a rhetorical activity, that planning could be seen as 'world making' (Fischler 1995). Not in the sense that plans and visions automatically determine a material and spatial outcome. Rather that such words, signs and symbols become the frame of mind for social agents as well as being the outcome of the historical and contextual conditions under which they are articulated. In other words, social agents 'appropriate' space through socio-spatial practices and identification processes.

Summing up, the cultural sociology of space framework is grounded in the dialectial socio-spatial relation in terms of material preconditions of actions, and in terms of their constitutive meanings (Sayer 2000). The basic proposition is that the socio-spatial relation *works* by means of its coercive or enabling capacities for spatial practices. Furthermore the socio-spatial relation conveys *meaning* to social agents via multiple representations, symbols, and discourses. Thus the socio-spatial relation on the one hand expresses possibilities and limitations to social actions within the built environment. On the other hand the meaning and valuation of this relation is constantly negotiated and re-negotiated on the basis of social imageries and cultural values. This dialectic tension furthermore expresses a politics of scale in the sense that socio-spatial practices and meanings produces and re-produces spatialities at scales from the body to the global, as in the case of the new forms of socio-spatial mobility.

The cultural sociology of space should be understood as a basic notion of how socio-spatial relations are marked by both practices and symbols. We see the cultural sociology of space as a (crude) framework for understanding all sorts of material practices and their symbolic meanings. However, if the particular object of study was for example the 'geography of the body' and everyday life practices, the cultural sociology of space might have been developed in a different way, and perhaps related to different research methodologies. Here, however, we are interested in understanding a particular new political agenda for ordering the European Union's territory. For that explicit analytical purpose we have developed a more policy targeted tool, namely that of analysing 'spatial policy discourses'.

## Analysing *spatial policy discourse*

The discourse analytical framework must be seen as an operational and analytical tool for dealing with the fact that spaces and places are represented in policy discourses in order to bring about certain changes in socio-spatial relations and preventing others. Policy discourses can be analysed in terms of their language, practices and power-rationalities (Richardson and Jensen 2000; Jensen and Richardson 2004). Thus we next explore these new dimensions of the spatial policy discourse of the European Union. First we analyse how representation of European space – of problems and opportunities for spatial policy – are framed in the new policy language. Then we consider the new spatial practices which allow the institutionalisation of the new discourse, before moving on to discuss the power rationalities which run through the new field of transnational policy making. In each case we attempt to draw out the framing of mobility and identity in the new policy discourse, making use of the conceptual framework to identify how the politics of scale and mobility are played out in the policy process.

*Language*

The first step is to explore how particular actions, institutions, or physical artifacts are re-presented in the language of policy documents. In the plethora of new European spatial policy processes, a series of documents charts the progress and shifts in policymaking, the emergence of new ideas, the entwining of strategies, policies and actions. Key documents are fragments of different knowledge framing processes. Thus different ways of framing 'space' set up different requirements for 'spatial knowledge' to be gathered and analysed in particular ways, to feed and support different spatial representations.

The core ESDP vision is expressed in what might be termed a new language of European spatial relations. It centres on a policy triangle of economic and social cohesion, sustainable development and balanced competitiveness. These objectives are to be pursued through the development of a balanced and polycentric urban system, new partnerships between urban and rural areas; securing parity of access to infrastructure and knowledge; and sustainable development, prudent management and protection of nature and cultural heritage (CSD 1999:11). Each of these terms carries particular meanings, but also leaves room for interpretation, having been first coined and shaped in a complex gestation process of policy development. Thus linking particular mobility rationales to spatial practice inhacing the accessibility as one of the crucial features of the European collaboration. The question of mobility has long been a vexed issue within the EU. Developing a common transport policy has been difficult, and subject to deeply entrenched disputes between member states (Whitelegg, 1992). What is indisputable, however, is that EU transport policy has always placed increasing mobility at its heart. Achieving 'sustainable mobility' was a key policy theme in the early 1990s, but within the ESDP the policy language has shifted. While the rhetoric of the ESDP returns frequently to the theme of mobility, the problem of mobility is framed in two ways. Firstly, particularly for regions on the periphery, as a problem of *accessibility*, and secondly, particularly for the core regions, as one of *efficiency*.

The ESDP states that improvements to accessibility are regarded as a critical priority in the development of the polycentric urban system and furthermore as preconditional in enabling European cities and regions to pursue economic development within an overall spatial strategy of harmonisation. Thus the notion of frictionless mobility and the cities as nodes in a polycentric spatial development model are two sides of the same coin:

> Urban centres and metropolises need to be efficiently linked to one another, to their respective hinterland and to the world economy. Efficient transport and adequate access to telecommunications are a basic prerequisite for strengthening the competitive situation of peripheral and less favoured regions and hence for the social and economic cohesion of the

EU. Transport and telecommunication opportunities are important factors in promoting polycentric development. ... Spatial differences in the EU cannot be reduced without a fundamental improvement of transport infrastructure and services to and within the regions where lack of access to transport and communications infrastructure restricts economic development (CSD 1999:26)

Within the ESDP, continuing an emphasis in EU documents from the Maastricht treaty to the Delors White Paper, the primary policy response is clear: the construction of trans-European transport networks (TENs) to remove barriers to communication and facilitate economic convergence and competition. TENs are identified as the area of existing EU spatial policy with most relevance to the ESDP process in terms of spatial development impacts and financial implications (CSD, 1999:14). Indeed, their development is regarded as crucial to the economic and social aims of the ESDP as well as potentially contributing to the third environmental objective. Furthermore, specific policy options, such as the 'dynamic zones of economic integration', are particularly dependent on infrastructure development. It is stated that policy measures in such areas, which could include the structural funds in Objective 1 areas, should focus on providing a 'highly efficient infrastructure at transnational, national and regional level' (CSD, 1999:21). Significantly, the ESDP states that prioritisation of development of the major arteries and corridors of the TEN-network will not suffice. It is necessary to upgrade the regional transport networks that will feed into the TEN, if economic benefits are to be secured. Here the ESDP repeats the blurring of earlier Commission policy documents (Richardson, 1995).

Europe 2000+ (CEC, 1994), placing TENs at the centre of a pan-European spatial planning framework, identified the problematic double role of assisting the creation of the single market whilst enabling balanced development of the Community as a whole. Significantly, it was recognised that TENs create a tension between these core territorial issues. While global competitiveness requires 'continuation, even acceleration of the implementation of large-scale TENs', this is accompanied by the risk 'of an increase in the imbalances in the Union', stemming from the 'strengthening of the centre to the detriment of the periphery' (among other factors) (CEC, 1994). Achieving balanced development and internal spatial competitiveness relies on avoiding the centralising impacts associated with TENs. So, TENs are clearly identified as threatening to drive a wedge between European global competitiveness and internal spatial competitiveness. Richardson (1995) has argued that EU discourse manages to avoid the difficult policy implications of this dilemma by making a series of assumptions about the effects of infrastructure development, allowing the impression that TENs can achieve such divergent policy objectives. Whilst the ESDP repeats concerns about 'pump' effects (where new high speed infrastructure removes resources from structurally weaker and peripheral regions) and 'tunnel' effects (where such areas are crossed without being connected) (CSD, 1999:26), all of the policy options identified pursue the general aim of improving accessibility as a generic response. Thus the ESDP's analysis of the problem of accessibility in the EU is straightforward:

Good accessibility of European regions improves not only their competitive position but also the competitiveness of Europe as a whole ... Islands, border areas and peripheral regions are generally less accessible than central regions and have to find specific solutions (CSD 1999:69).

The examples of Sweden and Finland, where regional airports link to European gateways, are quoted. The consequent risks are explicitly recognised, that:

> improved accessibility will expand the hinterlands of the economically stronger areas ... the newly accessible economies will have to compete against the large firms and the competitive services in these economically stronger areas ... competition may well benefit the stronger regions more than the newly accessible weaker ones (CSD 1999:70).

Yet the policy response is no more than to suggest that such infrastructure improvements need to be seen alongside other sectoral policies and integrated strategies. So, within the ESDP, mobility is framed as accessibility, and accessibility is framed in economic terms. This rhetorical construction appears to ignore the rather different ways that accessibility is being used in transport policy debates. In the UK, for example, accessibility is a key idea in the rapidly burgeoning debate about social exclusion. Here access to employment, services, leisure, etc. are considered to be important policy concerns. A further European spatial trend recognised in the ESDP is the growth of development corridors, where new development concentrates along transnational and cross-border corridors in already relatively urbanised areas.

The second core element of the discourse is efficiency. The problem of mobility framed here is the growth in road and air transport with resulting environmental and efficiency problems. Transport trends within the EU threaten to undermine progress towards sustainable development targets (CEC, 1996). The need to promote alternative modes is emphasised, but with several strong caveats:

> however this objective must be achieved without negative effects on the competitiveness of both the EU as a whole and its regions ... [and] nevertheless, both road traffic for passengers and freight will remain of great importance, especially for linking peripheral or sparsely populated regions (CSD 1999:28).

Similarly, while the potential for high speed rail is recognised as a competitor to air travel in the denser regions,

> in sparsely populated peripheral regions, particularly in insular locations, regional air transport including short-haul services has to be given priority (CSD 1999:28).

Here again there is the question of the extent of harmony between EU and national policy discourse. Drawing again from the UK example, where policy discourse has shifted towards demand management and integration, efficiency of networks is certainly an increasingly important objective. However, in the UK, policy has shifted away from road building, which remains the major component of overall spending on trans-European networks. Where the UK is in the process of creating local targets for road traffic management, no such shift is discernable at the EU level. Another example of the contradictory effects of the general mobility policy of efficiency can be drawn from the Danish case of the fixed link over the 'Great Belt'. After one year of operation, the number of cars crossing the bridge already exceeded the most optimistic forecasts made by the proponents of the bridge, undermining the official Danish policy of reducing car traffic. Elsewhere, in the accession countries, increasing road traffic levels – car ownership in particular – have been positively welcomed as signs of freedom in the post-Soviet era. Indeed, the rhetoric of the ESDP suggests that growth in overall traffic movements will be the key to improving accessibility.

At the end of the day, the ESDP relates to the ambiguous rhetorical device that one finds in the policy triangle of *growth-ecology-equity*, or what Eser terms the 'magic triangle' (Eser 1997:18). As the ESDP vision is biased towards the growth side of the 'triangle' it can be seen as an expression of the basic economic rationale that lies behind European co-operation. However an interesting point is that such a rationale is contradicted by the political integration philosophy based on the ideology of a common European culture and society. This not only contradicts the growing emphasis on competition between cities and regions. It also addresses the 'implicit rationale' of the ESDP. Thus, below the discussion about growth, ecology and equity lies the question of how to imagine a European territorially based identity and unity. This is certainly no coherent rationale as it more than anything else is coined around the idea of a 'Europe of difference'. Thus the process goes on from here, leading the ESDP discourse into yet another terrain of tension and conflict. So the explicit rationales of economic growth in terms of global urban competition, environmental sustainability and finally social cohesion (spatial justice), are underpinned by an implicit rationale of re-imagining European identity (Jensen 2001).

In contrast to the explicitness of the theme of mobility in the ESDP, revealing the discourse of identity requires more interpretation. However, we shall propose to understand the ESDP as more than a strategic *spatial policy discourse* focusing on the frictionless functionality of the European Union's territory. In our reading and interpretation of this phenomenon the ESDP can also be seen as an arena for re-articulating the notion of 'Europeanness', or in the words of John Urry:

> The Development of a possible 'European identity' cannot be discussed without considering how massive patterns of short-term mobility may be transforming dominant social identities (Urry 1995:169).

The basic rationale of argument is simply that mobility leads to cohesion that again (potentially) leads to an imagined identity.

*Practices*

Analysing key policy documents captures the representation of space in language, and reveals some of the power relations that contest these representations. This is, however, an inadequate analysis that needs to be placed within the context of a live policy process, where different interests compete for hegemony over the shape of policy, and where different spatial visions are contested. Spaces and places do not present themselves but are rather represented by means of power relations expressed in strategies, discourses and institutional settings. Although some of this is inherent in the text of policy documents, what is required is a broader view of the policy process that focuses on institutions, actions, and practices. So, in the multi-level processes of European spatial policy making, multi-dimensional conflicts inevitably arise. Here we simply outline the nature of some of these conflicts, from the invisible infranational workings of the European spatial policy community to the conflicts between regions and other interests who have a stake in both the visions and the implementation of EU spatial policy.

According to the analysis of the new policy language above, the ESDP is based on a process of re-imagining European territorial identity. However, the resulting visions are more than rhetorical devices, because they establish frameworks that direct EU measures towards a coordinated set of objectives. These new transnational processes are characterized by their dimensions of 'second order governance', informal procedures, networking and low transparency of the 'infranational level of governance' (Weiler 1999). In other words, the institutional and practice-oriented side of the discourse analysis makes the problem of democracy as transparency explicit. In this respect it is legitimate to voice concerns for the democratic perspectives of such new forms of planning. Furthermore, the ESDP process is an example of a new form of transnational planning where planners develop new capacities for networking and collaboration. They are creating new vocabularies for dealing with spatial issues, establishing such notions of space and knowledge as the 'natural way' of perceiving European space. The question of representing European identity and territory together with the rationales of growth, ecology and equity can be seen as part of a general theme of re-scaling that runs across EU policies (Brenner 1998).

The institution carrying the workload of writing and distributing the ESDP is the Committee on Spatial Development (CSD), established in 1991. The CSD has the unique status of being a transnational network of civil servants looking at European space from a new transnational perspective. It has thus become an institution that neither national nor EU politicians have full control over. This is not the least due to the 'infranational' character of the CSD. Infranationalism in the EU has been defined as the 'second-order Governance' involving commissions, directorates, committees, government departments, etc. These exhibit medium-to-low levels of institutionalisation, have the character of a network, practice an informal style and, last but not least, have a low actor- and event-visibility and process-transparency (Weiler 1999:275). These ways of working may imply a weakening of political control and increased autonomy being given to the administrative level, implying less control by the Member States and increased managerialism and reliance on expertise.

Alongside these new spatial practices relating to the ESDP's comitology, the question of implementation of the ESDP's policies has led to a further set of interesting practices. The first of these has been an emphasis on practical actions beyond the control of member states. Given that the ESDP has no legally binding status, it is not surprising that many of the possible actions identified within the ESDP focus on the transnational level (CSD 1999:35), thus avoiding the resistance of member states. Since 1996 the transnational INTERREG programmes have been the de facto field of implementation of the ESDP rationale and policies. Alongside this direct approach to implementation, the more subtle issue of the Europeanisation of state planning systems is inevitably raised by the existence of the ESDP, and the prospect of its gradual embedding in EU policy and law. As a further attempt to institutionalise and legitimise the ESDP's rationale, the document proposes that member states take its policy aims and options into account in framing national and regional spatial policies (CSD 1999:44). This is a very straightforward message to the member states that they should 'tune in' on this new Europeanisation of state, regional and urban planning in order to 'overcome any insular way of looking at their territory' (CSD 1999:45).

Further areas of competition exist between stakeholders, for example those representing urban and rural interests, between core and peripheral regions, and between cities and regions seeking to occupy key sites in the new polycentric map of European space. One of the most deeply seated tensions in European spatial planning is the divergence between the south European and north European view of the ESDP. Rusca identifies a characteristic of the distinctive 'southern attitude' as being particularly concerned with the cultural heritage and identity of places, for example (Rusca 1998). Mirroring this assertion of southern interests has been a move to articulate specific 'Nordic interests' in the ESDP. This is exacerbated by the prospect of EU enlargement, with the spatial vision extending into central Europe.

A further set of spatial practices relates to the reproduction of the new spatial policy discourse. In the words of Faludi, the success of the ESDP (for example) must be measured in terms of its ability to 'shape the minds' of social agents (Faludi 2001). As the emerging discourse becomes institutionalised in new spatial practices, including those of spatial analysis, the construction of knowledge forms and fields of knowledge results in boundaries being established between valid and in-valid, reasonable and unreasonable forms of knowledge. Such boundaries are vital in institutionalising European spatial planning as a 'rational, science based policy field', and at the same time they act as powerful instruments in the process of marginalising and excluding other forms of knowledge (e.g. radical environmental considerations or indicators of social equality). As an example of this 'knowledge policy' the recent appointed 'Study Programme on European Spatial Development' (SPESP 2000) should be emphasised. Apart from stressing a need for more comparable data and more solid knowledge of the spatial development of the European territory the document introduces an interesting new concept that illustrates the power-knowledge dimension. Thus in pursuit of a deeper spatial understanding the process must be supplemented with 'infography' (SPESP 2000:13). Behind this new concept lies a very explicit acknowledgement of the importance of spatial representations that takes a deliberate turn away from 'realistic' description. In recognising the rhetorical and powerful importance of spatial representations it is said that in recent years, numerous symbolic representations of the European territory have been created. Often have they presented mind-catching illustrations, which have served as powerful tools for both shaping attitudes and visualising policy aims. Some images even have become policy icons (SPESP 2000:13). In the ESDP, infographics are used to articulate the concept of polycentricity.

*Power rationalities*

Discourses frame and represent spaces and places, and thus express a specific power-rationality configuration. In the 'classic' sociological tradition, rationality is understood as the underlying structure of values and norms that governs social actions

(Weber 1978). However, rationality is inseparable from power (Flyvbjerg 1998). Construed in a productive way, power is here seen as the foundation for social action as well as potential control and coercion (Foucault 1990). Thus different rationales – with their distinctive horizons of values and norms that guide social actions – are implicitly acts of power in that they are attempts to govern what sort of social actions are to be carried out, and what are not.

Returning to our object of empirical study, the new spatial policy discourses of the European Union, the fusion of the cultural sociology of space and the analytical framework for analysing spatial policy discourses brings one of the embedded rationales out very clearly. In Hajer's words this is a spatial representation primarily in terms of the Europe of Flows (Hajer 2000:135). In a policy discourse of a Europe of Flows, regions and cities will increasingly present their visions as being repositioned and connected to the spatiality of flows. Thus this competition-oriented rationale is by far the most predominant of the nested rationales residing within the discourse. However, we shall also address the rationales of sustainable development, social cohesion and European identity building.

Our analysis suggests that as the discourse of the Europe of Flows is articulated and embedded in practice, its 'other' is marginalised. The hegemony of the Europe of Flows can only be understood as a mobility-, competition- and growth-oriented discourse that derives its distinctive identity in opposition to a Europe of places. In other words the dialectical tension between the two 'spatial logics' is represented in a distinct way in order to draw out the rationale for perceiving European spatiality in terms of flows and mobility rather than its opposite. Ascribing hegemony to the Europe of Flows by coining and re-presenting European spatiality in the vocabulary of flows is an act of 'naturalising' the increased urge to be a key player in global economic competition. Overall the Europe of Flows discourse thus enhances the notion of a multi-speed Europe in which different 'Europes' are superimposed on one another (Hajer 2000:144). Furthermore, such a notion clearly contradicts the idea of infrastructure enabling balanced development.

The rationale of sustainable development is thus subsumed into the logic of global competitiveness emanating from the notion of the Europe of Flows. This is in spite of severe problems with traffic congestion in the core region, which the ESDP recognises as running contrary to the core sustainability objective. This is also the case with social cohesion policy, which is the institutionalised hallmark of the European Commission's Directorate General for Regional Policy (DGXVI), which authored the ESDP. EU regional policy is underpinned by a vision of Europe as a spatially coherent unity. Cohesion policy aims to diminish regional differences, leading the way to a less fragmented territory. The rationale of competition maintains a hegemony over the goals of sustainability and cohesion in the ESDP, by being framed as the precondition for their attainment. Finally, we suggest that the whole complex of rationales that we have identified within the emerging spatial policy discourse underpin the 'imagined community' of Europe, thus contributing to a framing of European identity (Jensen 2001).

A specific example of this effect is found in the plan to use the ESDP as the foundation for a standard textbook on European geography for secondary schools across Europe (Ministers Responsible for Spatial Planning and Urban/Regional Policy 1999). This is a very clear expression of how this discourse attempts to condition the identities of social agents by framing specific ways of thinking about European space. Furthermore, it is an example of the micro processes of reproduction of spatial discourse. The ESDP might be thought of in terms of a new geographic imagination facilitating the construction of a particular European identity, and thus complementing the other new symbols of European Union unity – the flag, the hymn, the passport, and the single currency (Hedetoft 1997:152–3; Kohli 2000:121). In this sense the underlying discourse of European identity is akin to building an 'imagined community' (Anderson 1991).

## Framing mobility and identity

The aim of this chapter has been to show the benefit of bringing cultural sociology to bear on the analysis of spatial policy discourses. Such an approach might seem obvious or unnecessary, but it is part of our critical agenda to suggest an alternative to a-spatial policy analysis of spatial policy discourses. This rests on the assumption that discourses might be seen as social constructions but also that such spatial policy discourses dealing with representations of space must be understood in relation to their spatial 'object'. This is in no way an inclination to a simple correspondence notion of discourse and material/spatial reality. Rather we see the importance of understanding the discourses of space against the background of the cultural sociology of space that offers the meta-theoretical understanding of the relation between social life and its material surroundings. Thus we advocate a perspective of the cultural sociology of space, taking departure in an understanding of socio-spatial relations, as both a question of material constraints and enabling capacities as well as a realm of symbolic meanings and representations on spatial scales from the body to the global.

From the examples of European spatial policy, we have illustrated how socio-spatial relations, seen from the perspective of a combined framework of the cultural sociology of space and discourse analysis should include the languages, practices and power-rationalities of these policy discourses. Thus the words, images and languages used to represent and frame European space found in the ESDP reflect particular spatial symbolic meanings and representations, central to which is the contestation of mobility. Furthermore, such representations of space are a reflection of contemporary material globalisation processes which create the incentive for the European Union to facilitate action in new policy fields such as transnational planning, and policy making under the INTERREG program. Ultimately, such spatial policy discourses carry a (not necessarily coherent) mix of rationalities. Three competing rationalities of economic competitiveness, environmental sustainability and social equity surface in the ESDP. Furthermore, one could also interpret these new spatial policy discourses as contributing to a general discourse of European integration through the implicit notions of European community and identity. There is consensus amongst a number of analysts about viewing European identity as something that could be constructed, but not in opposition to national and sub-national identities. According to such thinking European identity cannot take its cues from national identity. Rather, what is needed is a 'new hybrid' (Kohli 2000:114). According to this view European identity is not opposed to national identity since identity should be understood as multi-layered (Kohli 2000:126) and implying a shared loyalty towards more than the (so far) hegemonic nation state level (Weiler 1999:346).

From our analysis the possibility of a theoretical and analytical perspective hinges on perceiving how the spatiality of social life is played out in a dialectical tension between material practices and symbolic meanings at scales from the body to the global. Thus any spatial policy discourse seeking to direct or produce new spatial practices works by means of constructing and reproducing new language uses and other practices expressing specific power-rationalities. So if the ESDP means anything, it both creates the conditions for a new set of spatial practices which shape European space, at the same time as it creates a new system of meaning about that space – based on the language and ideas of hyper mobility and polycentricity. Or as we have framed it elsewhere:

> Missing from the multiple narratives of Europe is a critical analysis of the relationship of this contested European project with the *spaces* of Europe. What does the political and economic project of European integration mean for its cities, its environment, and for its territory? In this book, we argue that a core dimension of the European project is the making of a single European space – *a monotopia* – and that this spatial agenda is becoming more explicit, and more coherent, with the advent of a new field of European spatial policy, which is embedding new ideas about relationships across space in a multi-level, transnational field of activity ... the term monotopia captures the idea of a one dimensional (mono) discourse of space and territory (topia/topos). It is our aim here to reveal the discourse of 'Europe as monotopia' as an organising set of ideas that looks upon the European Union territory within a single overarching rationality of making a 'one space', made possible by seamless networks enabling frictionless mobility ... a space of *monotopia*. By this we mean an organised, ordered and totalised space of zero-friction and seamless logistic flows ... The future of places and people across Europe seems closely linked to the possibility of monotopia (Jensen and Richardson 2004:ix–3)

This analysis of spatial policy discourses has identified significant implications for the identity building process, even though the question of cultural identity is not explicit in the text of the ESDP (Dutch National Spatial Planning Agency 2000:70). Furthermore, this question will gain even more weight with the enlargement of the European Union. Here this sort of thinking could work as a stimulus to the new real or imagined geographies of Europe. In this process such representations of space and their accompanying transnational institution- and network-building might function as a vehicle for further European integration. Although the explicit discourses underlying the ESDP are mutually challenging, this analysis suggests an interpretation of the implict discourse as one of building European identity. Thus at the end of the day we see the adding of a layer of 'Community building' to the notions of identification with the 'European idea', or 'constitutional patriotism' in Habermas' words (1996:507). The point to observe here is that the new spatial discourse complements this 'loyalty to the Idea' with an attempt to articulate a 'loyalty to the Territory'. The very important question remaining is whether this opens the political agenda for using documents like the ESDP in an act of identity construction coined around openness and a 'progressive sense of place' (Massey 1993), or whether the effect is rather the opposite.

# References

Anderson, B. (1991) *Imagined Communities. Reflections on the origin and spread of Nationalism*, London: Verso.

Bauman, Z. (1998) *Globalization: The Human Consequences*, Cambridge: Polity Press.

Bauman, Z. (2000) *Liquid Modernity*, Cambridge: Polity Press.

Beauregard, R. A. (1995) 'If only the City could speak. The Politics of Representation', in Liggett, H. and D. C. Perry (eds) (1995) *Spatial Practices. Critical Explorations in Social/ Spatial Theory*, London: Sage, pp. 59–80.

Beckmann, J. (2001) 'Heavy Traffic Paradoxes of a modernity mobility nexus', in Nielsen, L. D. and H. H. Oldrup (eds) (2001) *Mobility and Transport – an anthology*, Copenhagen: The Danish Transport Council, pp. 17–25.

Berman, M. (1983) *All That Is Solid Melts into Air – The Experience of Modernity*, New York: Verso.

Brenner, N. (1998) 'Between fixity and motion: accumulation, territorial organization and the historical geography of spatial scales' *Environment and Planning D*, 16 pp. 459–481.

Castells, M. (1996) *The Information Age: Economy, Society and Culture, vol. I: The Rise of the Network Society*, Oxford: Blackwell Publishers.

Castells, M. (1997) *The Information Age: Economy, Society and Culture, vol. II: The Power of Identity*, Oxford: Blackwell Publishers.

CEC (1994) *Europe 2000+: Co-operation for European territorial development*. Luxembourg: Office for Official Publications of the European Communities.

Commission of the European Communities (CEC) (1996) *Towards sustainability: Progress report from the Commission on the implementation of the European Community programme of policy and action in relation to the environment and sustainable development*, COM(95) 624 final, 10.1.96. Brussels: CEC.

CSD (1999) *European Spatial Development Perspective – Towards Balanced and Sustainable Development of the Territory of the EU*, Presented at the Informal Meeting of Ministers Responsible for Spatial Planning of the Member States of the European Union, Potsdam May 10/11 1999, CSD.

Dutch National Spatial Planning Agency (2000) *Spatial perspectives in Europe*. Ministry of Housing, Spatial Planning and the Environment, The Hague.

Eser, T. W. (1997) *The Implementation of the European Spatial Development Policy: Potential or Burden?* Trier: TAURUS Diskussionspapir Nr. 1.

Faludi, A. (2001) 'The application of the European Spatial Developmeny Perspective: evidence from the North-West Metropolitan Area', *European Planning Studies*, 9: 667–679.

Fischler, R. (1995) 'Strategy and History in professional Practice: Planning as world making', in Liggett and Perry (eds) (1995): *Spatial Practices*, London: Sage, pp. 13–58.

Flyvbjerg, B. (1998) *Rationality and Power. Democracy in Practice*, Chicago: The University of Chicago Press.

Foucault, M. (1990) *The History of Sexuality Volume 1: An Introduction*, London: Penguin.

Graham, S. and S. Marvin (2001) *Splintering Urbanism. Networked infrastructures, technological mobilities and the urban condition*, London: Routledge.

Habermas, J. (1996) 'Citizenship and National Identity', in Habermas, J. (1996): *Between Facts and Norms. Contributions to a Discourse Theory of Law and Democracy*, Cambridge: Polity Press, pp. 491–515.

Hajer, M. (2000) 'Transnational networks as transnational policy discourse: some observations on the politics of spatial development in Europe', in Salet, W. and A. Faludi (eds) *The Revival of Strategic Spatial Planning*, Amsterdam: The Royal Netherlands Academy of Arts and Sciences, pp. 135–142.

Harvey, D. (1996) *Justice, Nature and the Geography of Difference*, Oxford: Blackwell.

Harvey, D. (2000) *Spaces of Hope*, Edinburgh: Edinburgh University Press.
Hedetoft, U. (1997) 'The Cultural Semiotics of "European Identity": Between National Sentiment and the Transnational Perspective', in Landau, A. and R. Withman (eds) (1997): *Rethinking the European Union. Institutions, Interests and Identities*, London: Macmillan Press, pp. 147–171.
Jensen, O. B. (2001) *Imagining European Identity. Discourses Underlying the European Spatial Development Perspective*, Paper for *the Lincoln Institute* conference on European Spatial Planning (ESP), Lincoln Institute, Cambridge MA, 29/30 June 2001.
Jensen, O. B. and T. Richardson (2001) 'Nested Visions: New rationalities of Space in European Spatial Planning', *Regional Studies*, Vol. 35.8, pp. 703–717.
Jensen, O. B. and T. Richardson (2004) *Making European Space. Mobility, Power and Territorial Identity*, London: Routledge.
Kohli, M. (2000) 'The Battlegrounds of European Identity'. *European Societies* 2(2): 113–137.
Lefebvre, H. (1974/91) *The Production of Space*, Oxford: Blackwell.
Massey, D. (1993) 'Power geometry and a progressive sense of place', in Bird, J. et al. (eds) (1993): *Mapping the Futures. Local Cultures, Global Change*, London: Routledge, pp. 59–69.
Miller, P. and Rose, N. (1993) 'Governing economic life', in Gane, M. and Johnson, T. (eds) *Foucault's new domains*, London: Routledge, pp. 75–105.
Ministers Responsible for Spatial Planning and Urban/Regional Policy (1999) *ESDP Action Programme*, Final Version 18.9.99.
Perry, D. C. (1995) 'Making space: Planning as a mode of thought', in Liggett, H. and D. C. Perry (eds) *Spatial Practices. Critical Explorations in Social/Spatial Theory*, London: Sage, pp. 209–241.
Richardson, T. (1995) 'Trans-European Networks: good news or bad for peripheral regions', proceedings of *Pan-European Transport Issues Seminar*, pp. 99–110, PTRC 23rd European Transport Forum.
Richardson, T. and Jensen, O. B. (2000) 'Discourses of Mobility and Polycentric Development: A contested view of European Spatial Planning', *European Planning Studies* 8 (4) pp. 503–520.
Ritzer, G. (2001) *Explorations in the Sociology of Consumption. Fast Food, Credit Cards and Casinos*, London: Sage.
Rusca, R. (1998) 'The Development of a European Spatial Planning Policy', in C. Bengs and K. Bohme (eds) *The progress of European Spatial Planning*, Stockholm: Nordregio (1), pp. 35–48.
Sayer, A. (2000) *Realism and social science*. London: Sage.
Sennett, R. (1990) *The Conscience of the Eye – The Design and Social Life of Cities*, London: Faber and Faber.
Shields, R. (1991) *Places on the Margin. Alternative Geographies of Modernity*, London: Routledge.
Smith, N. (1993) 'Homeless/global: Scaling places', in J. Bird, B. Curtis, T. Putnam, G. Robertson, L. Tickner (eds) *Mapping the Futures. Local cultures, global change*, London, Routledge, pp. 87–119.
SPESP (2000) *Study Programme on European Spatial Planning*, Draft Final Report 3 March 2000, to be found at http://www.nordregio.a.se.
Thrift, N. (1996) *Spatial Formations*, London: Sage.
Urry, J. (1995) *Consuming Places*, London: Routledge.
Urry, J. (2000) *Sociology beyond Societies. Mobilities for the Twenty-first Century*, London, Routledge.
Weber, M. (1978) *Economy and Society*, Berkeley, University of California Press.
Weiler, J. H. H. (1999) *The Constitution of Europe. "Do the New Clothes have an Emperor?" and other essays on European*, Cambridge, Cambridge University Press.
Whitelegg, J. (1992) *Sustainable Transport – The Case for Europe*. London: Belhaven.

# Chapter 6

# 'Your Passport Please!' On Territoriality and the Fate of the Nation-State

Wolfgang Zierhofer

## Introduction

What happens when we show our passport? We constitute and reproduce a social, political and territorial order that is both local and global – the modern political sphere of the world-society which is dominated by the institutional form of the 'nation-state'. In this text the passport serves as a point of departure for the analysis of those practices that at the same time establish the local and global aspects of political subjects, identities, rights and interactions – i.e. it illustrates the constitution of the political sphere of the world-society.

The institution of the 'nation-state' is designed to provide coordinates for the prescriptive location of activities, persons, organizations, objects and other entities on a global scale. However, the claim of the inter-state system to serve as an exclusive frame of reference is illusory. Organizations are contingent structures, and in principle every single one of their functions or services could be offered by other organizations. Indeed, increasingly often we encounter forms of governance where states and institutions of civil society pursue projects jointly. Moreover, a considerable corpus of literature has emerged dealing the question of the future of the nation-state and trans- or post-national constellations. In particular, processes of economic globalization and deterritorial allocations of power reveal the limits of political representation within a global system of territorial states.

We may easily conceive of a non-modern political sphere which, by reflecting upon the self-constitution of the political sphere, avoids the naturalization of political institutions and communities that is so characteristic of modernity. However, there are no discernible signs of institutions or practices that could replace even the modest guarantees of legal and political recognition that the modern nation-state provides internally and on the global scale. In this respect the possible dissolution of the interstate system and a general 'outsourcing' of political rights, legal guarantees and social security may raise serious ethical concerns. At present at least, a bunch of customer, credit and client cards does not seem to be a viable substitute for your passport!

## The practice of passport control

'Your Passport Please!' – we all know this ritual, but what kind of product is established by these practices? Certainly your identity is checked in order to control international mobility. But there is more to it: this routine act articulates a global network of social relations which finally serves to put people, activities and many other things in *their proper place*. Controlling passports means setting up a modern kind of 'Aristotelian space': A normative frame of reference is applied in order to evaluate the actual locations of people and many other things, and, if necessary, to redirect them to the prescribed position. Within this act a specific *world order* is articulated.

With an analysis of passport checks I attempt to reveal two main things: the level of discursive *constitution* and reproduction of our social world, and, in particular, the specific character that this takes on in late modernity. In so doing I offer an understanding of the world-society that conceives of the interstate system on the one side and the individual subject on the other hand as entities that are constituted simultaneously by the same practices. This perspective takes the world-society as a network of practices that constitute one other and all other known entities. To constitute an entity does not mean to realize it or to give it existence, but rather to differentiate it from other entities by articulating its specific identity in relation to others. Subsequently I will focus on the nation-state – the form that dominates the organization of the world-society – and elaborate on the conditions of its future significance. Approaching the story through passport controls is heuristically motivated: given the network character of social practices, we could unfold a perspective by starting with virtually any activity.

Alternatively, we could also plunge directly into the extensive debate about globalization and the future of the nation-state. But what we would then miss is precisely this unity of micro-practices on the one hand and global structures on the other that is so nicely illustrated by passport controls. Actually, all practices are both micro and macro, local and global – in a specific way and by necessity. But passport controls deserve our attention because they belong to those activities that establish the dominant form of order. As I will argue, this order is not the outcome of a scalar world, as many other texts implicitly assume, but on the contrary the precondition of scale. It provides a frame of reference that permits us to distinguish between locales and between varieties of local and superlocal or global levels. The following three levels are of basic importance:

1   *The individual.* The act of control classifies persons and constitutes central aspects of their identity. It creates a legal and political subject and relates it to a community and its institutional structures: Persons are localized in a social space. They are regarded as specific citizens, and this determines the conditions of their international mobility. The same act also constitutes the border guard and the whole border infrastructure as a representation of a particular country. These practices of surveillance constitute a particular class of places.
2   *The nation-state.* By confirming a national identity, the passport control reproduces relations between subjects and nations, between different nations and between subjects of different nations. The identity control locates the traveller and the official in terms of their territorial affiliation and their physical location.

Goods, other items and even environmental effects that pass a border also have to undergo similar controls. These controls do not necessarily have to take place at the physical border, they can be carried out virtually anywhere. As a precondition, this whole complex system relies mainly on the previous territorial definition of nation, state, citizenship, property and general localization. So, people and other entities get a location in a space which has a variety of social and physical dimensions.

3   *The interstate system.* According to the logic that is enacted at customs and border checks, whatever crosses a border belongs to a state, which has a certain responsibility for it. Passport controls only make only sense if identity documents are edited by organizations that mutually acknowledge their right and their ability to do so. The territorial state and its sovereignty exist only as a form within the inter-state system. However, this order did not become global and universal until the twentieth century. In earlier times there were spots on the world map that did not represent nation-states, and even 'outlaws', i.e. persons with reduced legal protection and undefined territorial affiliation. Today there are only 'in-laws' on our planet, and the worst that can happen is to be without papers. According to the logic of the inter-state system there is no intentional possibility, except as an administrative failure, of being without a national identity.

Summing up, we may say that the inter-state system constitutes subjects, nation-states and the global inter-state system at the same time. This act of constitution is more than a formality on paper: it does not define subjects, nations and a specific world-society in the sense that an encyclopedia would refer to these entities. Rather, these entities only exist in the form of passport controls and related actions. Editing birth certificates, registering people as inhabitants of communities, locating them within a bio-social kinship system, classifying them in racial, sexual and ethnical terms and other acts of this kind, establish systems of inclusion and exclusion, systems of access to opportunities to form one's own life. What we individuals are and how we live have evolved within such contexts.

The world-society, its institutions, individual biographies, experiences and bodies are all constituted and shaped by the same actions. There is hardly any escape from this 'system' and we can grant it the character of a total institution. Nevertheless, it neither determines individuals nor the society. Rather, it is a situational disposition, which provides individuals with specific opportunities to move to other situations and 'write' their biographies.

Within this all-encompassing location system the social space is of primary importance, and physical space is only subsidiary. Physical conditions, particularly the locations of human bodies and other things, are not ends in themselves, but instruments for achieving and maintaining a certain social order. In the end, all social qualities have to be translated into physical differences in order to achieve permanence and in order to be reproducible. This requires either codes of corporeality (e.g. sex, race) or arbitrary, but physically reproducible codes (e.g. names, numbers, classifications).

Interactions have to be meaningful und legitimate in order to be reproduced. This is usually achieved by the discursive reflectivity that actually constitutes the interaction. Just imagine starting an ethnomethodological experiment at a border:

| Immigration officer: | Your passport please! |
| Passenger: | What do you mean by passport? |
| Immigration officer: | Your identity documents, please. |
| Passenger: | Sorry, but I have a fluid identity. |

From this moment on the officer will start to work through the steps of a programme designed for refractory individuals, constantly offering ways back to normality. And thus the meaning and legitimacy of passports, citizens, nation-states and the global inter-state system are secured – if necessary with sanctions, even physical violence.

### The 'metaphysics' of the modern inter-state system

Rituals accompany many activities that take us from one life context into another. In his *Rites of Passage* van Gennep (1986) explains the rituals that accompany the transfer of a person or a thing from one spatial realm or from one episode of life into another because the community needs to maintain order during this process of transformation. There are ceremonies that provide spiritual support for the great passages in life like birth, marriage and death, but also for membership of institutions like a church, a school, the army or an enterprise. The French 'adieu' and the German 'Grüssgott' remind us that our greeting rituals, whether we encounter a person in the street or enter a house, still reproduce the form, but for most actors with a different significance.

Nevertheless, rituals as they are executed nowadays at border checkpoints also establish transcendental references in some ways. They still accompany the transition of a person from one community or power-container to another one by referring to a security system. Nowadays, however, a secular context has replaced the metaphysical ones of earlier times. A legally defined institutional framework of surveillance and sanctioning is regarded as sufficient to secure specific socio-political orders in particular communities. It is useless to appeal to spiritual forces in order to ban evil. Although exclusively human actors establish this secular security system – it operates beyond the awareness of the average subject. Thus it also largely escapes political negotiation and achieves a quasi-transcendental and naturalistic status – like the nation-state itself.

It is the same discourse or network of practices that gives meaning to passport control and shapes and dominates the constitution of the political sphere of modernity, particularly in its formal and institutional aspects. Probably the central category in this discourse is the nation-state. Nation-states are brought into existence as elements of a system of co-coordinated actions. The elements and the encompassing system are not only mutually constitutive, they exist in the form of the very same activities. In this sense the nation-state and the inter-state system are only two sides of the same coin (Taylor 1994, 1995). However, it is not enough to conceive of the nation-state as a territorial unit and container. In respect of social relations, too, we have to regard it as an element that is constituted by a system, which is itself recursively, constituted by this very element. This second system, which actually encompasses the territorial inter-state system, is the political sphere of modernity (Taylor 1999, pp. 69ff). A basic function of the nation-state is to create and maintain order at many levels and in many respects. It establishes a social 'space' (Williams and Smith 1983), which provides the 'coordinates' for 'locating' all kinds of activities, persons and things, and directing them to their proper 'place' – if necessary by violence. Since it coordinates many other institutions and practices too, the nation-state is truly a meta-institution.

In the following I will first turn to some of the various ways in which the nation-state localizes items on a global scale. In the subsequent section, however, I will turn the perspective upside down and interpret the nation-state as an institution that globalizes practices. Passports, as certificates of citizenship, localize their 'subjects' as well as providing specific preconditions for their global mobility. The local and the global are not naturally given differences, but distinctions that stem from certain practices. Other practices would probably provide us with other instruments of reference and other ways to establish differences in scale. But the practices that establish the inter-state system and the corresponding spaces and scales have become the dominant frame of reference in contemporary debates on globalization and world orders. Despite the growing critical awareness of 'state-centred' approaches in political theory, many scholars still follow this modern metaphysics, and thus miss chances to draw practices into the realm of political evaluation and negotiation. Symptomatically, the contemporary transformation of the modern political sphere has usually been approached in terms of issues like the 'retreat' or even the 'end of the nation state' (Guéhenno 2000) or the advent of a 'post-national constellation' (Habermas 1998, and similar terms are used by Albrow 1998, Beck 1997). The fact that the nation-state constitutes itself as an element of a global system is on the one hand a precondition of its capacity to unfold its ordering power. On the other hand, it bears within it the seeds of a process that might, in the long run, overthrow this specific order. In sections five and six I will argue that we are most likely witnessing the first signs of a transformation of the political sphere into a non-modern sphere.

## The state as global localizer

The nation-state is the modern successor of former orders and forms of rule that were primarily based on kinship, charisma and religious legitimization (Opello and Rosow 1999). Although sovereignty has been steadily depersonalized and secularized since the Middle Ages, the nation-state, succeeding the feudal and the early territorial state, has maintained the character of an overall framework. Although government has been split up in a balance of power and although personal power has been transformed into institutions and collective procedures, it has remained centralized: Ideally only one government was entitled to determine the fate of the territorial unit called a state – a term which referred originally to the personal 'condition' of the prince (Skinner 1997) and subsequently remained attached to less personalized forms of rule and government. It is only recently that weak forms of institutionally heterogeneous government, usually labelled 'governance', have gained official recognition and legitimacy. This might mark a turning point in the history of the organization of the political sphere.

*The state as container? – Territoriality, community, culture*

Within modernity the nation-state claims exclusive political representation and sovereignty for itself. It is an institutional network that determines the proper position within this network for all human activities. Above all, the state is designed to contain legislation, which subsequently defines the forms, scopes and conditions of accepted activities, whether they are simple interactions or organizations. Freedom is not located outside this order, but has its place assigned internally, either as a pre-defined variation of regulated activities, or else as clearly demarcated unregulated regions, much like white spots on a map. The location of activities is not primarily determined within coordinates of physical space, but within social, institutional spaces. Nevertheless, this often involves corporeal criteria and territorializations.

By definition, nation-states represent communities and vice-versa. The nature of such communities was and still is contested. Primarily cultural or ethnical interpretations of nations, as popularized by Herder and Fichte (Schulze 1999, pp. 170ff) and by many nineteenth century geographers (Schultz 1998, 2000) have in the meantime been scientifically discredited. Nevertheless, nationalism of this sort is still a major cause of discrimination, conflicts and bloodshed. Political theory today conceives nations as imagined communities (Anderson 1998), which are indeed constituted by experiences within institutional frameworks, above all those of the territorial state. In the technological conditions of pre-modern societies the identification of communities with territories seemed to be fairly straightforward, even to most geographers, as Werlen (1999, pp. 13f) remarks. But there were always non-territorial communities, such as gypsies and religious communities.

At least in Europe, the conceptual link between community and territory seems to stem from agricultural power systems, which depend for their fiscal reproduction on a controllable link between fertile soil and a human workforce. This meant an immobilization of individuals, at least as long as labour markets had not yet been developed – these were only stimulated by the industrial 'need' for mobile means of production. Before modernity, states and empires achieved neither a high degree of territorial independence nor a significant level of real-time operability across large distances (Giddens 1985, pp. 49ff). Only with the development of aircraft and electric communication technologies has it become possible to realize distance-independent, and thus non-territorial, placeless communities, which operate as organizations in real time for all their individual members. Thanks to the telephone and radio broadcasting, individuals could for the first time in history communicate about their lives with virtually any individual or even all others. This had tremendous consequences for the imagination of the national community *and* of the world in general.

Modern nation-states constitute their communities though legal membership – inscribed on the body and embodied by the passport – and territoriality is above all a vehicle for surveillance and for the accumulation and allocation of resources. If there is still an intimate relationship with culture, it is provided by the state, which subordinates individuals to specific rules of conduct. By *socializing subjects* the nation-state turns itself into a cultural container. Its institutional framework not only provides similar experiences for its subjects; it also unites them in a *community of common destiny*. In particular, politically participating citizens' experience of sharing in the intentional (re)formation of the nation and the state – in the sense used by

Seyes and Renan (Schulze 1999, pp. 168f) – may be an important factor in national identification, cohesion and willingness to share collective forms of behaviour. While regarding modern states as cultural containers, we must avoid the conception of homogeneous culture. Rather, national culture, like an official language, is only a specific set of behavioural dispositions among others, upon which individuals rely to coordinate their activities in certain contexts.

## A universal system of coordinates and addresses

According to Luhmann (1998) the territorial state serves as a 'mould' for the internal segmentary differentiation of the political system of the world-society. This matrix-like entity splits the global political sphere up into very different segments of equal kinds. On the one hand they provide the possibility of adapting politics to local circumstances such as resources, culture, infrastructure etc. On the other hand the nation-state has become the general form of address. States tend to accept only other states as partners, and they prefer them as enemies to other organizations. The 'community' of states secures a world order that is dominated by reliable, or at least calculable 'actors' and related standards of interaction. Moreover, the nation-state is also the major reference for the functional differentiation of the political sphere, particularly on the two scales that are defined by the internal and external space of the nation-state. The inter-state system not only establishes a global social and territorial order; by the same token it also establishes series of similarly structured domestic orders.

Every state is an economic system: it has to control the flow and distribution of resources in order to concentrate power. Collecting taxes provides the means to keep the executive body of the state operable. By localizing values and competencies, states create an internal administrative-economic space. Because all corporeal entities have some exclusive extension in physical space and time, all forms of government, rule and control over physical resources (minerals, animals, plants, workforce) and resources which are bound to physical entities (information, experience) involve some kind of territoriality. Territoriality establishes access to and accountability in respect of the kind of corporeal entity concerned. Usually, access to resources, including strategic locations like harbours or passes, but also the flow of the means of production and of goods, are socially and territorially regulated. Supported by forms of mercantilist politics, these aspects constitute relatively autonomous 'national economies'. In combination with the state sector (taxes, expenditures) these politico-economic spaces establish communities of common experience and fate at the aggregate level.

Although armed force is an indispensable resource for the reproduction of a state, it is primarily legitimacy that guarantees stability and permanence. The latter is derived not only from historical narratives, but above all from actual living conditions. In order to build and reproduce infrastructures for internal and external security, for health, social security, education, mobility, communication and so on, the state has to play a major structuring role within the domestic economic space. In the end, these services and functions are essential in ensuring the loyalty of citizens.

*Scale and identity*

The most obvious aspect of state territoriality is the continuity of segmentary differentiation from the inter-state system down the scale to an inter-community system. These organizational settings form common structures of individual lifeworlds and contexts of socialization. Furthermore, they provide a standard environment for organizations and a range of opportunities to set up parallel institutional structures, for instance in the sports sector, and to assign territorial identities on different scales.

Nation-states legitimate themselves and their form of rule by referring to a nation – that is, to a *natural* and therefore quasi-transcendental community. Within nation-states *persons* are, by implication, constituted as individual citizens, who in turn represent the nation and vice versa. This mutual representation is organized along territorial and social dimensions. The inter-state system constitutes and enforces the rights of citizens on a global scale, which means that it also constitutes aliens, displaced persons or 'illegals', but also render literal 'outlaws' impossible. By the logic of its construction, the inter-state system is a total institution. No one is left beyond accountability. This is an aspect not only of surveillance, but also of legitimacy, because any internationally accepted identity provides access to specific resources, services and opportunities. Furthermore, it defines a mutual set of rights and duties shared by the state, its institutions and the individual. A similar socio-spatial locating and accountability applies also to enterprises, associations and other sorts of organizations, as well as to things, whether they are valuable resources, information or just problematic waste. With few exceptions, they belong to some state and they have to behave according to certain laws.

While the inter-state system constitutes only *national* citizens, in its interior the modern state constitutes subjects along many dimensions. It thus creates social positions such as officials and private persons, soldiers or civilians, politicians or voters, taxpayers, pupils, judges, professionals with an accredited diploma, migrants, social security beneficiaries and so on. Similar structures apply to organizations and objects. In principle, all regulative differences that are introduced by some of the institutions of the state provide a dimension of the identity and the lifeworld of subjects and other entities. However, they are not all equally important.

Because the nation-state defines itself as the political representation of its citizens, it constitutes individuals as subjects in the double sense of autonomous actors who have the ability to enjoy rights and have duties on the one hand, and individuals who are structurally subordinated to and physically incorporated in the territorial state on the other hand. Along the scalar institutional order, identities too mutate from the local or communal to the international level.

These modes of constituting subjects are not mechanical processes, like the stamping of materials. Rather, they take place as discursive differentiations within interactions, and as related situational settings which provide typical experiences, and, in the end, typical biographies. The constitution of subjects is therefore both the defining of positions and their significance, as well as the *socializing* of individuals along pre-defined lines.

*The state, the political sphere and contingency*

While the form of the nation-state seems to have stabilized, if not congealed, the political sphere in general is exposed to considerable transformations, of which the growing independence of state-defined and territorially confined politics is probably the most remarkable. Although all self-constitutions of states rely on naturalizing rhetoric (Taylor 1999, p. 78), the state is nothing but a human invention, and thus a *contested concept* that serves as a blueprint for contingent structures. States are sluggish organizations, which nevertheless need some flexibility in order to claim their position in the world of competing organizations. There will always be groups and organizations that try to reduce or replace certain functions of the state, extend the acceptable definition of states, present themselves as complementary or overarching organizations (like the UN, the EU or NGOs), revolutionize the existing order or challenge the state's claim to exclusivity in other ways.

But interestingly enough, the strongest challenge to a state-centred world order stems from internal inadequacies or even contradictions. The very dual structure of the local state on the one hand and the global inter-state system on the other establishes a problem which I would like to call the *paradox of territoriality*. By extending the form 'nation-state' on a global scale, this system constitutes worldwide subjects of the same kind. From a legal point of view, human beings all over the world are in principle the same, not just undefined *others* that could be regarded as slaves or as some other minor form of existence. They are all entitled to a passport. This remarkable historical fact could be a major source of diffuse cosmopolitan orientations which, at the end of the day, will challenge all territorially limited systems of political representation. Intended to establish a general system of localization, this order globalizes forms of recognition that are in the end incompatible with forms of territorially confined political, moral and legal recognition.

**The state as local globalizer**

As an element of the inter-state system, the state is the end and access point of a global network. The inter-state system provides a formal uniformity, a code for political addresses. In so doing it establishes worldwide conditions of mutual recognition for individuals and institutions, which in turn provide a global system of accountability for persons, things, activities and their consequences. For the world-society the state guarantees specific forms and degrees of operability, reliability and effectiveness. Just as the modern state offers global standards for localizing activities and corporeal entities, it offers all these locations specific types of access to global institutional structures. And just as the standards embodied by passports ensure certain rights and procedures in respect of persons worldwide, other standards fulfil similar mobilizing functions for other entities and certain activities. The inter-state system is an enabling precondition of globalization.

Although worldwide trade is much older than the territorial state, it is the establishment of a global system of nation-states at the end of the colonial era (Opello and Rosow 1999, pp. 181ff) that provides the opportunity to extend the realm of economic activities over which the state need not have immediate control. In the first instance the global inter-state system provides nothing more than a channel of relatively undistorted political communication. This, however, is a favorable condition for enhancing the conformity of many fields of activity. In respect of the mobility of activities, particularly industrial production, this means a general reduction of transaction costs. And, in addition, mercantilist policies can be confined on the one hand to securing a few strategic domestic advantages, like licences or patents, while on the other hand supporting the general opening and extension of markets. Without assuming a historical teleology, we can interpret this process as the continuity of the modern process of rationalization and attempts to exploit marginal advantages.

With Habermas (1981, p. 28) we might then ask: According to what kind of rationality? No doubt, many problematic aspects of economic globalization could easily be interpreted as consequences of instrumental rationalizations, as a further extension of the 'system' and a colonialization of the 'lifeworld'. However, this image is not complete before an increase in terms of communicative rationality is also taken into account. The chances that people around the world will get their claims heard and considered in decision-making have increased considerably in the course of the last few decades.

In respect of the state we could speak of two analytically different, though interwoven processes of globalization: on the one hand, the institutions of the state and their representatives work with global standards for procedures and conditions which ease the mobility of capital, resources, production sites and to some degree also of workers. The state thereby creates favourable conditions for global markets in goods, services and capital (Jessop 1999). On the other hand, however, the inter-state system universalizes the modern form of the political subject, thus realizing the unity of the political world-society. More or less at the same time, several technological achievements, such as the telephone, radio, TV, the Internet, nuclear weapons and environmental risks, but also tourism, migration and internationally operating NGOs have fostered awareness of a de-facto world-society not only among politicians, but also among citizens. Both processes, the rather active *politico-economic globalization* and the more passive *cosmopolitization of horizons* seem to bear the seeds of a repositioning of the territorial state and the inter-state system within the global political sphere. For both processes the state functions as a scalar switch, as a global localizer and local globalizer.

Over the last decade a flood of literature has been devoted to the repositioning of the nation-state. Developing this aspect of globalization by focusing on the state inevitably leads to certain kinds of 'territorial traps'. This is to be avoided by taking the state in the first instance as an organization, and thus a *social* structure, and by paying attention to the transformations of the political sphere of the world-society. However, this requires a certain analytical distance from the mythological and metaphysical self-assurances of the state. One has to take into account that the state, despite its modern monopoly position, is only a contingent, constantly changing and historical organization.

According to the modern understanding of the republican state, this organization was simultaneously able to represent the interests of capital and work. Democracy would ideally provide the possibility to transform class conflicts into class compromises. In conditions with high mobility of goods, services, capital and production sites, the state seems to have lost its capacity effectively to represent the interests of workers, employees, and its citizens in general. Compared with capital and machines, human beings and above all ecological conditions are highly immobile (Bauman 1999). Profit accumulates globally, but most negative effects remain local (Beck 1997, pp. 100ff.). While the resources of the state are still excellently suited to the liberalization of economic activities, it has lost its grip on many business activities, and in consequence has become increasingly unable to control and steer their consequences according to the political will of its citizens. The whole security system of the territorial state was designed to save its population from attacks by other territorial states and from threats from other citizens within the same state. Not only globalized risks like ecological threats and non-territorial enemies, such as terrorist networks, but also the very means of securing national welfare, such as international competition policies, call the legitimacy of the state-related structures of political representation into question.

By the logic of its construction, however, the inter-state system on the one hand establishes global spaces to localize people, things and activities, and on the other hand a global recognition of political subjects. The latter is a precondition for the development of an effective cosmopolitan sphere. And that is precisely what is fostered by the global institutional weakness of the nation-state. More and more people begin to see their lives as interrelated with events taking place elsewhere on the planet. On the basis of their traditional understanding of political participation, they want to express their will so it has an influence on the government of their community. But in the light of dense global networks of interaction and influence, the promise of the nation-state to transform their collective intention is no longer so convincing. The citizens may ask themselves: What is our formal recognition worth if our interests are not taken into account where it really counts?

As in earlier ages, the forms of political representation seem to have reached a state of oppressive contradiction. The contemporary problem, however, is not related to the extension of political and legal recognition to other classes of human beings, like slaves or women, but to varieties of territorial inclusion and exclusion that are intrinsic to the modern structure of the political sphere.

## The modern constitution of the political sphere

According to the modern self-conception of the state, it provides the institutional framework for all 'political' activities, either as a formalized container for varieties of activity, or as a programme for building networks of organizations, or by setting normative limits to realms of unregulated interaction. This position as meta-institution is legitimized by modern meta-narratives of rationality, equality, emancipation etc. and by the principle of political representation of individual citizens. The strength of this conception is revealed, when, for instance, a critical mind like Ulrich Beck talks without any irony of 'sub-politics' (1993, pp. 154ff) when referring to decisions that have considerable consequences, but are not explicitly made within 'political' institutions.

In order to mark a clear difference between a modern and a non-modern constitution of the political sphere, I will draw on Bruno Latour's interpretation of modernity. For Latour (1995) the moderns believe they are able to separate the world into nature and culture, without intending to leave a position for anything of a third kind. This is indicated in Figure 6.1 by the vertical bar. The moderns now refer to precisely this insight in order to mark the difference between modernity and pre-modernity, as indicated by the horizontal bar. According to their view, non-modern cultures are unable to distinguish consistently between nature and culture and to handle these conceptions in a rational manner. They live in worlds of hybrid entities. As a consequence, they can neither develop science and technology on the one hand, nor positive law, democratic political systems and complex institutional frameworks on the other hand.

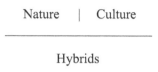

Nature    |    Culture


Hybrids

**Figure 6.1    The constitution of modernity**
According to Latour (1995, p. 20).


Latour uses this scheme primarily to criticize the implicit ascription of a transcendental status to nature and culture. He and many other contemporary sociologists of science argue rather that the distinction between nature and culture is contingent, and that the meaning of these two notions is constantly being produced and reproduced. Moreover, it is constituted precisely by the hybrid entities under the bar. This is best illustrated by the scientific laboratory, which serves to fuse things, instruments, processes, technologies, energy, information and human beings into an experimental setting, which will only through its outcomes – and thus afterwards – reveal a component that is nature and a component that is culture. Hybrid settings are created in order to draw a distinction between natural laws on the one hand and cultural techniques on the other.

What Latour was aiming for with this scheme was more than just a treatment of nature and culture. I tend to interpret it as a critique of the dominance of binary and transcendental thinking in modernity. Expressed in different semantics, this critique has become standard in postmodern and poststructuralist approaches. With Gotthard Günter (Klagenfurt 1995) we may also read it as an attempt to overcome the classical (Aristotelian) logic that knows only two values. It is a plea for the recognition of third positions or for the excluded other (Figure 6.2).

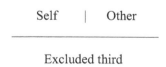

Self    |    Other


Excluded third

**Figure 6.2    The modern constitution of relations of recognition**

If we re-interpret Latour's scheme as a representation of relations of recognition and non-recognition, we get a key to the modern structure of the political sphere. Indeed, we saw that the inter-state system universalized the recognition of human beings as political subjects. Yet Beck's notion of sub-politics reminds us that there is a realm that is not recognized as political in the proper sense (Figure 6.3). The binary distinction between the Self and the Other represents the modern sphere of political recognition: politics is about conflicting interests, persuasive speeches and presenting better arguments, and the subjects are constituted as political antagonists or even enemies. This is the world above the horizontal bar. Below we find technological developments, business decisions, cumulative outcomes of market operations, institutional transformations and so on, which are not recognized as political processes *per se*, but rather as 'private' activities.

My/our political interests      |      Political interest of the other/s

Non-political sphere, subpolitics

**Figure 6.3   The constitution of the modern political sphere**

A historical perspective, however, would inform us that the determination of the political, the political sphere and the political community were always contested. Since the ancient Greeks the sphere of political recognition has been extended to slaves, women, foreigners and other categories of human beings. However, we must not regard this extension as a smooth linear development, since not only the status of citizens, children, insane people and criminals, but also the legal recognition of animals for instance (Ferry, 1992, pp. 9ff) has changed several times and in different directions. Both the idea of the community and the status of the excluded third may vary considerably. Modern thinking, however, takes its conception of the community for granted. This is most obvious in nationalist accounts of nations as natural communities (sometimes even with natural boundaries), and, interestingly enough, in the semantics of immigration offices, where foreigners become 'naturalized' when they are granted full political rights.

Within the modern political sphere, the exclusion of classes of activities and, as a consequence, of parts of the world from political representation is not constituted as a political issue. Through its constitution the political naturalizes itself and its organizing principles. Only by systematically disregarding the constitutive preconditions of politics could the nation-state achieve a quasi-transcendental status within modernity. Otherwise, it would have been acknowledged as a very important, though contingent organization *beside* others. But the modern state claims exclusivity (there is only 'the' state), universality (in the end the state determines everything) and globality (the world consists of states). For a modern mind the state represents the political community and the organizing principle of the political sphere. It seems misplaced to allow people from other countries to participate in national elections or referenda, and it would be absurd to grant animals, machines or other non-human entities political representation.

But perhaps politics and political representation should not be thought of in terms of modern institutional structures either. In spite of all formal non-recognition, many excluded thirds are in fact already politically represented in one way or another. They enter the parliaments through the back door as elements of discourses, and they are often given a voice in the form of arguments.

## Some characteristics of a non-modern political sphere

At the abstract level, a non-modern constitution of the political sphere accepts first the profound contingency of all constitutive differences, and secondly the realm of all those possible thirds that are not represented within politics, but are constitutive for the distinction between the political and its environment. The borderline of the political sphere is itself acknowledged as a political issue. Political self-constitution is therefore not naturalized, but *taken into consideration*.

This implies that no entity, no aspect of the world is a priori and irrevocably excluded from political representation. However, it does not imply the granting to all non-citizens of the same kinds of political rights. Nevertheless, the notion of sub-politics becomes superficial because there are no natural or transcendental boundaries of the political. The first act of political self-recognition is to consider and legitimize a contingent organizational boundary of political representation. There is no 'sub-' outside politics, only a procedural delimitation of institutional settings.

In consequence, any naturalistic understanding of 'nation' is rejected: people need neither belong *a priori* nor for their entire life to a particular political community, nor is the structure of political representation within a community confined to its own members. Rather, the pro-active recognition of a world-society is reflected in worldwide structures of representation, although decisions always have to be made within a specific organizational framework. It is at least conceivable that any problem can be articulated at any node in communicative networks. The concept of a local Agenda 21, which focuses on the responsibility of municipalities for global environmental and social conditions, may serve as an illustration of the general principle.

It is not the acknowledgment of a world-society and global networks of interaction that challenges the position and legitimacy of the modern nation-state. Such notions are older than the modern nation-state, and they were probably even a reason for the establishment of a quite standardized international political sphere. What is problematic is rather the territorial limitations of political representation, which deprive the state of some of its powers. The territorial nation-state is actually based on a *categorical fallacy*, because it takes populations of administrative regions for political communities.

Moreover, there is no stable substance to the state. Our world-society is populated by many sorts of mobile entities, a fact which tends to delegitimize the concept of the nation-state as a container for and exclusive exponent of these entities. Subjects, citizens, inhabitants, goods, capital, and production sites – they all come and go. What remains is nothing but a temporal form of self-constitution, which constantly has to be filled by flows of entities of all kinds in order to remain present. The second problem of the modern nation-state is therefore that the nation is taken as a natural, territorial, political community, whereas the state should be regarded as a contingent historical organization. As if were not enough that it is a categorical fallacy, it has even been naturalized.

In the course of the second half of the twentieth century the dominant position of the state and the exclusivity of the inter-state system became precarious. From a non-modern perspective the state is not necessarily regarded as a universal form, but rather as an important organization beside others. This opens up scope for different forms of governance on all scales. From the point of view of the individual, who is interested in effective political representation, this is an ambivalent issue. Certainly, social movements and related NGOs may fight for certain ideas or for the interests of their members on a global scale. But are they able to provide all human beings with formal guarantees of political participation and equal representation?

Moreover, a shift from the monopoly of the nation-state to a plurality of political organizations would provoke a qualitative transformation of forms of representation. The relevance of elections and voting might wane in favour of arguments, lobbying, direct action, the economy of the mass media and the buying of 'representation services' on global markets. In principle, all functions and services of the state could be provided by other kinds of organizations, which would then replace the one nation with many specific, though overlapping communities. Observing and imagining 'gated communities', we can sense some of the problematical consequences. We must not forget that within the institutional framework of the nation-state individual rights and political representation were linked to individual duties, above all taxation. If this framework is broken up, the *general principles* of the redistribution of burdens and benefits, particularly all ideals of equal treatment irrespective of origin, age, gender etc., will also be at risk. This may raise considerable ethical questions. What would it mean to replace the passport with an individually tailored set of credit, customer and membership cards? What kind of communities, what kind of territorialities, what kind of scales would emerge? So far, neither political theory in general nor political geography have taken these developments systematically into account.

A non-modern understanding of the political sphere thus neither presupposes any political community, nor does it *a priori* exclude any entities from political communities. Rather, *politics is conceived as the communication that aims to reach understanding over forms of co-existence*. This is of course a much broader field than what the moderns have in mind. The idea of structuring political communication by establishing an overarching organization is doomed to failure because possible alternatives are ignored. As soon as the nation-state and parliamentary democracy are regarded as a sub-field of politics, however, the possibilities of competing organizations and the historical relativity of the nation-state are acknowledged. But this does not imply a need to abandon the nation-state. On the contrary, such an approach might provide scope for a critical and argumentative evaluation of the nation-state and its relation to political communication in general, as well as to other possible political institutions. This might be a chance to save the transformation of the global political sphere from losing itself in 'sub-politics' and pull it into the realm of negotiations for political recognition.

## Conclusion

By approaching the subject through practices related to passports, this chapter has portrayed the modern nation-state as the key element of the global inter-state system. The latter has two aspects: the localizing aspect consists of global social and physical spaces, through which the inter-state system constitutes nation-states, their internal orders and their subjects. Its globalizing aspect consists of local but uniform institutions and procedures which constitute a worldwide sphere of political recognition. The nation-state is the form that differentiates the world-society into equivalent segments which serve in many ways as links between the global and the local, and thus as scalar switches.

In the light of the progressive integration of the world-society, the limitations of a territorially structured political sphere have become obvious. The state is no longer regarded as the only, inevitable container of political activities. Other organizations, such as enterprises and NGOs, compete for influence on all scales, and people all over the world develop a cosmopolitan awareness. But is this also the end of the nation-state, as several authors would have it? We certainly cannot predict its future. But by reflecting the role of the nation-state within a non-modern political sphere, we can distinguish those aspects that are actually challenged or discredited from those that have not been called into question so far.

It is not the many localizing functions of the territorial state that have become problematic, but its claim to provide sufficient structures for political representation within a self-conscious world-society. Over the last few decades organizations of various kinds have established global structures of political representation that complement the nation-state to some degree. It seems very likely that the state will also have to 'outsource' many other functions and services. Nevertheless, so far there is no sign of an institution that might replace the nation-state as a universal political address, nor are there prototypes of alternative institutions that could *guarantee* basic protection and political representation on a global scale, irrespective of the status and the origin of the subject.

But the idea of granting basic recognition and support to all human beings is of course a modern one. Who knows what postmodernity is going to do with it?

## References

Albrow, Martin (1998): *Abschied vom Nationalstaat. Staat und Gesellschaft im Globalen Zeitalter*. Suhrkamp, Frankfurt a.M.

Anderson, Benedict (1998): *Die Erfindung der Nation. Zur Karriere eines folgenreichen Konzepts*. Ullstein, Berlin.

Bauman, Zygmut (1999): 'Local Orders, Global Chaos'. In: *Geographische Revue*, No. 1, pp. 64–72.

Beck, Ulrich (1993): *Die Erfindung des Politischen*. Suhrkamp, Frankfurt a.M.

Beck, Ulrich (1997): *Was ist Globalisierung?* Suhrkamp, Frankfurt a.M.

Ferry, Luc (1992): *Le nouvel ordre écologique*. Grasset, Paris.

Gennep, Arnold van (1986): *Übergangsriten*. Campus, Frankfurt a.M.

Giddens, Anthony (1985): *The Nation-state and Violence*. Polity Press, Cambridge.

Guéhenno, Jean-Marie (2000): *The End of the Nation-State*. University of Minnesota Press, Minneapolis.

Habermas, Jürgen (1981): *Theorie des kommunikativen Handelns*. Suhrkamp, Frankfurt a.M.

Habermas, Jürgen (1998): 'Die postnationale Konstellation und die Zukunft der Demokratie'. In: Habermas: *Die postnationale Konstellation. Politische Essays*, pp. 91–169. Suhrkamp, Frankfurt a.M.

Jessop, Bob (1999): *Globalization and the National State* (draft), published by the Department of Sociology, Lancaster University at: http://www.comp.lancaster.ac.uk/sociology/soc012rj.html.

Klagenfurt, Kurt (1995): *Technologische Zivilisation und transklassische Logik*. Suhrkamp, Frankfurt a.M.

Latour, Bruno (1995): *Wir sind nie modern gewesen*. Akademie, Berlin.

Luhmann, Niklas (1997): *Die Gesellschaft der Gesellschaft*. Suhrkamp, Frankfurt a.M.

Luhmann, Niklas (1998): 'Der Staat des politischen Systems. Geschichte und Stellung in der Weltgesellschaft'. In: Beck, Ulrich (Hrsg.): *Perspektiven der Weltgesellschaft*, Suhrkamp, Frankfurt a.M, pp. 345–380.

Opello, Jr. Walter C.; Rosow, Stephen J. (1999): *The Nation-State and Global Order. A Historical Introduction to Contemporary Politics*. Lynne Rienner, Boulder and London.

Schultz, Hans-Dietrich (1998): 'Deutsches Land – deutsches Volk. Die Nation als geographisches Konstrukt'. In: *Berichte zur deutschen Landeskunde*, Vol. 72, No. 2, pp. 85–114.

Schultz, Hans-Dietrich (2000): 'Land – Volk – Staat. Der geografische Anteil an der "Erfindung" der Nation'. In: *Geschichte in Wissenschaft und Unterricht*, Vol. 51, No. 1, pp. 4–16.

Schulze, Hagen (1999): *Staat und Nation in der europäischen Geschichte*. Beck, München.

Skinner, Quentin (1997): 'The State'. In: Goodin, Robert; Pettit, Philip (eds): *Contemporary Political Philosophy*. Blackwell, Oxford. Pp. 3–26. [Originally published in *Political Innovation and Conceptual Change*, ed. T. Ball, J. Farr and R. L. Hanson, Cambridge University Press, 1989, pp. 90–131.]

Taylor, Peter J. (1994): 'The state as container: territoriality in the modern world-system'. In: *Progress in Human Geography*, Vol. 18, No. 2, pp. 151–162.

Taylor, Peter J. (1995): 'Beyond containers: internationality, interstateness, interterritoriality'. In: *Progress in Human Geography*, Vol. 19, No. 1, pp. 1–15.

Taylor, Peter J. (1999): *Modernities. A Geohistorical Interpretation*. Polity Press, Cambridge.

Werlen, Benno (1999): 'Regionalism and Political Society'. In: Embree, Lester (Hrsg.): *Schutzian Social Science*, pp. 1–22. Kluwer, Dordrecht.

Williams, Colin and Smith, Anthony D. (1983): The national construction of social space, *Progress in Human Geography*, Vol. 7, No. 4, pp. 502–518.

Chapter 7

# Territoriality and Mobility – Coping in Nordic Peripheries

Jørgen Ole Bærenholdt

## Introduction

In his historical novel *Dramar á jördu* (2000) / *Drømme på jorden* (2001) / *Dreams on Earth*, the Icelandic author Einar Már Gudmundsson uses as a motto a quote from Halldór Laxness (*Frie mænd*) on the role of mobility and distance in a poor twentieth century Icelandic family (translated from Danish):

> They were seven, and they disappeared to remote places: two sons drowned in remote seas, one son and one daughter disappeared to an even more remote country, America, even farther away than death, maybe. However, no distance is greater than the one that separates poor relatives in the same country.

In times and spaces of poverty in the Nordic peripheries, mobility was a matter of separation. Unlike the rich and powerful Odysseus, few came back. Even poor relatives on remote farms in the very same parish were distant from one another. The paradoxical element in Laxness' (Gudmundsson's) motto, 'no distance is greater than the one that separates poor relatives in the same country', relies on the implicit territorial norm that distances should be shorter within the same country.

Distances are material, social, and cultural. It has been a key challenge for Nordic nation-states and municipalities (local states) in the twentieth century to overcome these distances in the Nordic peripheries. Infrastructure development, welfare policies and the promotion of 'the people's' cultural unity have gone hand in hand in the significant modernization policies that have been applied to the Nordic peripheries. These policies coped with peripherality not just by overcoming distances; they involved the whole socio-spatial constitution of modernity in places, to the extent that one can question whether 'periphery' is any longer the correct label. In addition to the role of states, municipalities have played a key role as 'transformers' in this process.

The history of the socio-spatial practices of people living along the Nordic Atlantic coastal rim of Norway, the Faroes, Iceland and Greenland in the twentieth century includes remarkable cases of empowerment and embedded entrepreneurship – cases of coping with poverty and distance. Facing the mobilities of the twenty-first century, these Nordic Atlantic practices are now challenged by non-territorial entrepreneurialism and the growth of tourism. One reason for looking at the Nordic Atlantic peripheries here is that the experiences of rural peripheries may have more to contribute to social theory than we normally think, leading us to question and reconsider the dominant images of rural peripheries (Therborn, 1999; Cloke and Little, 1997).

First an approach to territoriality, mobility and coping is presented. Then the three dimensions of coping – innovation, networking and the formation of identity – are discussed in general for the Nordic Atlantic peripheries in three sections. Finally, cases of coping from the municipalities of Ilulissat in Greenland and Hornafjördur in Iceland are studied. The conclusion discusses the challenges of coping among inhabitants of the Nordic 'peripheries' in the twenty-first century.

## Territorial and mobile coping – the approach

If social relations and space are to be approached as questions of production, and thus not as something pre-ordained, we must look into the practices and processes that produce social relations and space. Indeed the 'and' between social relations and space needs clarification: Societies are produced with certain spatialities; space is not an external feature of societies. The same argument goes for time and temporalities, although this is not the focus of this paper. Space, like time, is more than an external feature of social relations and societies. Whether we approach societies as territorial states or as face-to-face social relations, spatial organization, spatial processes or simply spatial practices are constitutive dimensions. Hence, coping with distance has probably been constitutive in the socio-spatial formation of most societies, and this is certainly apparent in the case of the Nordic peripheries.

The social relations and spatialities of human life, in the Nordic peripheries as elsewhere, can be approached as a coping process whereby people try to master the limited possibilities of social construction that make sense to them. Like many other concepts, coping is a concept of socio-spatial practice, and it involves the 'productive' use of power relations. Here territoriality and mobility should be approached as complementary dimensions of socio-spatial practices, where the use of boundaries is combined with the use of movement. Therefore, we investigate '... explicitly the process of the territorialization of space, the construction and signification of demarcations and boundaries' (Paasi, 1996:7–8). However, coping is not only a question of *territoriality* (Sach, 1986) but also of *mobility*, defined as an incorporated *social strategy of influence and control through the movement of human beings, things or information* (Bærenholdt, 2001:142; Bærenholdt and Aarsæther, 2002:157).

The approach is constructivist-materialist, and this means that the production and power of space intersects with materialities, social-spatial practices and spatial discourses. As such, the multiple productions and powers of Nordic Atlantic (as well as other) spaces are about nature, landscapes, economy, culture and politics, but a cross-cutting centre of these processes is the social-spatial practices of people; practices that I call coping.

The coping approach has been developed for the UNESCO MOST (Management of Social Transformations) Circumpolar Coping Processes Project.[1] In locality studies in Nordic countries as well as Russia and Canada, we have used the (normally psychological) concept of coping as a concept of social practice. Coping (or 'coping *strategy*') can be defined as a set of practices in three dimensions (economic, social and cultural):

1   *Innovation*: The process of change in economic structures resulting from new solutions to local problems, as responses to the transformations of a globalizing and increasingly knowledge-based economy;
2   *Networking*: The development of interpersonal relations that transcend the limits of institutionalized social fields;
3   *Formation of identities*: The active formation of identities that can reflect on cultural discourses from the local to the global (Aarsæther and Bærenholdt, 2001:23; Bærenholdt and Aarsæther, 2002:153).

In some respects the approach involves a modernist normative stance favouring a correlation between innovation, networking and the formation of identity. It includes a pragmatic 'bottom-up' stress on a crucial nexus between the meaningfulness of socio-economic development and the productivity of the socio-cultural discourse mediated by socio-political networking.

All three dimensions are spatial. Innovative restructuring involves spatial restructuring of localization, connections and movements of activities. Networking relations mean specific forms of proximities between actors. And identities are formed with more or less explicit references to geographical imaginations. Space is neither a container nor simply an outcome of these practices. Practices, including coping strategies, are spatial, and all practices exercise power in one way or another.

Territorial and mobile practices are two principal ways of exercising power spatially. However, the exercise of power through mobility and territoriality is deeply interconnected (Bærenholdt, 2001). If one is to control and defend a territorial area of settlement, and harvest natural resources to live on, mobility is an indispensable technique for controlling borders when there are not enough walls and guards to keep permanent control. On the other hand the territorial control of humans and information would be of little relevance if the use of means of transport and communications did not make mobility in everyday life an issue.

'Coping' signifies an understanding of the production of social relations and spatiality where human practices struggle in creative ways that are neither total mastery nor mere adaptation. People cope with challenges and conditions over which they do not have control. On the other hand, they are not just victims of developments and conditions. Coping is somewhere between a 'strategy' and a 'tactic' (de Certeau, 1984), because coping can neither be demarcated in relation to the other, nor does it fully belong to the other. Coping is about how people manage to make a relatively secure life for themselves in a world that they cannot fully master, but where mere adaptation is not enough.

Spatiality is 'inside' this creative struggle; spatial entities cannot be taken for granted. For example, there is a trend towards community-based territorialization, but such communities cannot be taken for granted. They may be involved in EU policies that bypass states, thereby spatially producing new territorial demarcations and identifications (Storey, 2001:139–142). Hence, communities are always in a process of spatial definition, and such definitions may also involve mobile networks or flows.

Spatialities are produced, and they are not only territorial. Wealth is a question of mobility – of 'dromocratic power' (Virilio, 1998). Few technological innovations have been as crucial to historical development as those within transport (Braudel, 1981). Coping is also very much a question of mobility. For a long time, the territorial character of embeddedness and empowerment has not been questioned, but the tacit role of territorial principles of social organization is now being challenged by an increasing awareness of the socially constitutive meaning of mobility in late modernity (Urry, 2000; Bauman, 2000).

Building on the discussion of mobility and social capital in Putnam's analysis of the crisis of 'civicness' in the US (Putnam, 2000), we can consider his distinction between bonding and bridging social capital. While the notion of social capital has been associated with types of bonding social relations and norms of trust somehow embedded in specific regions, others have pointed to the non-embedded character of productive social relations. Putnam therefore admits that bridging can be as important as bonding in dynamic social relations.

Where Putnam uses bridging and bonding as adjectives for social capital, I see bridging and bonding as an approach to the practices of coping strategies. Bridging is a concept for coping that builds connections between different social groups or fields, while bonding characterizes the practice of binding one social group or field together as opposed to others. In other words, bridging establishes new, cross-cutting social connections, while bonding defends and intensifies already established social groupings.

We should not confuse bonding-bridging with territoriality-mobility. Bridging is not mobility and bonding is not territoriality; such an understanding would only lead to seeing the bonding of social and cultural groups as a question of spatial borders. Mobile bonding communities are indeed on the twenty-first century agenda (Urry, 2000), while the question of territorial bridging of groups and individuals as citizens of a political territory has been at the centre of Nordic Atlantic strategies. Territorial bridging is often the critical aspect when it comes to bridging different villages, social groups and cultural communities at the municipal level. The socio-spatiality of coping can be analysed in the continua of both bridging-bonding and territoriality-mobility (see Figure 7.1). The following three sections discuss the socio-spatialities of coping in the Nordic Atlantic area in the 1990s within each of the three dimensions: innovation, networking and the formation of identity.

**Figure 7.1   Socio-spatialities of coping**
*Source*: Bærenholdt and Aarsæther, 2002:162.

## Innovation: combining mobility and territoriality

Fisheries have been crucial to livelihood in the majority of Nordic Atlantic localities in the twentieth century. While neither fluctuating fish stocks nor international fish markets can really be mastered, the innovations introduced in response to 'outside' transformations are emblematic cases of coping. This is also true in the growing new tourism economies; tourists are not easier to catch and control than fish. Tourists may come on their own initiative, and this can even happen more than once, but they also tend to escape easier than fish.

Because of the highly mobile and international character of fisheries and the fish trade, and the relatively dispersed character of settlements in the Nordic Atlantic, people have been used to finding local solutions as responses to problems of a non-local character for centuries. Coping with increasingly knowledge-based economies is a further challenge, where questions of brain drain and attracting the educated back are on the agenda. In this case the development of means of mobility (bodily, virtual or imaginative, Urry 2000) plays a crucial role in innovation.

Innovations in economies of mobile fish and tourists are often about the spatial organization of flows more than the 'processing' itself. The value of the fish to the consumer or of the experience to the tourist is to a great extent a matter of 'being in the right time and place'. Since spatial framing and movement tend to be the 'production' as such, combinations of different types of territoriality and mobility are crucial aspects of business innovation.

Another basic innovation involved in Nordic Atlantic strategies is their innovations in social organization. In economic geography, there has been an interest in explaining the economic performance of firms embedded in associational economies (Cooke and Morgan, 1996), and it seems that associational capacity is not only a matter of firms but of social organization in a broader sense.

Cases of innovation in Faroese, Iceland and Greenlandic fisheries[2] always combine elements of informal networking, municipal entrepreneurship and the ability of mobile professionals to commit themselves to local development. In one Faroese case, the buildings for a new internationally oriented firm in a small village were constructed by means of voluntary work in the village; which is said to be the normal 'Faroese way' (Hovgaard, 2002). In Icelandic and Greenlandic cases, municipal authorities played crucial roles as facilitators for new business developments. Links to international markets are often in the hands of a few businessmen who in some way commit themselves to local development. In the end, much Nordic Atlantic experience points to the importance of 'territorial capital'. While the control of mobile marine resources is, as far as possible, a matter of national schemes for resource management, local actors who retain control of financial capital locally are likely to be powerful enough (have enough political capital) to influence resource allocation policies in their own interests.

Nordic Atlantic resource-based economies, fisheries, tourism or the like, have often been socially organized in combinations of mobile bridging and territorial bonding. Territorial bonding is used to defend and manage natural resources, while mobile bridging organizes contacts with international markets. The organization of such contacts is mobile, since the power is exercised by means of movement; and it is bridging, since new social relations are established in marketing and sales. Co-operative and municipal organization have been used to enforce local and regional economic regulation in order to secure stability and in some cases distributive justice.

Innovations in the Nordic Atlantic are less about Schumpeterian technology development than about social innovations for dealing with and adapting to the dynamic complexities of natural resources and international markets that cannot be controlled. These are innovations promoting *regulation* and stability in conditions of disorder. The spatial organization of social relations is the focus of such innovations. When much of the 'production' concerns the careful transport of fish and tourists, mobile bonding with those people 'from here in other places' (diasporas) and mobile bridging with new business partners are crucial. However, associational organizations for voluntary work, territorial control of financial capital and territorial commitment of professionals are all about territorial bridging among different social and cultural groups in terms of some kind of local citizenship. Territorial definitions of micro-societies are thus not taken for granted but produced as part of their social construction. Territorial practices such as the commitment of the diasporas to their 'place of birth', and the innovative capacity of commitments to do something for the inhabitants in one's place of birth, are due to the combination of mobile bridging with other places and territorial bonding with the 'place of birth'. If you cannot visit, communicate or transport across distances, committed actions by outsiders are of little value. Still, the innovative capacities of such 'mobiles from here' would be of little use in producing local societies if the process were not regulated by principles of territorial regulation. New business interventions by mobile capital with no territorial commitment in fisheries and tourism present challenges to the 'traditional' social innovations (associational organization, territorial control) familiar from the twentieth century.

### Networking: easy bridging versus normative bonding

Since networks are in a sense the crucial innovations, the distinction between innovation and networking is not absolute. This point is further evident from the study of the development in time and space of specific local projects. Coping has much to do with making connections and securing access, and it can be facilitated by powerful municipal authorities and other local actors taking part in multi-level governance relations in regional development and policy (Jessop, 2000; Bærenholdt and Aarsæther, 2001).

The socio-spatiality of networking is to a great extent about the production of certain kinds of local authorities, an area where we can see differences in the scale and degree of formal institutionalization, with the Faroes and Norway as the extremes. Networking in Norwegian cases has much to do with ways of coping with national regimes of institutional regulation, so that for example everybody gets a share of the pie of secure local public-sector service jobs. This is for example the case in the Storfjord municipality in Northern Troms County, in contrast to Faroese cases, where crucial differences are more due to social entrepreneurship in the business development of cornerstone companies (Bærenholdt and Haldrup, 2003; Aarsæther and Bærenholdt, 2001).

Networking is *normative*, and much net*working* is about working. In Nordic countries, there has been an ongoing production of a sense of community (*Gemeinschaft*) based on the participation of everyone in work, historically based on seventeenth century Pietism: 'To see that everybody works is the main principle of Nordic societal organization and the force that holds associations together' (Stenius, 1997:164).

Whether networking is territorial or mobile, the work in question is also about the control and definition of deviance. So far we have considered bonding networks; it is bonding networks that are normative. The central feature of *bonding* is that it clearly defines who is included and who is excluded. As such, bonding networks are not open-ended. Independently of their spatial organization, bonding networks work as a result of a distinction between who are 'insiders' and who are 'outsiders'.

In bridging networking, directions and routes can easily be reoriented to exclude some and include others in the network at work. Bonding networking is the opposite; exclusion from bonding networks is harder to enforce. Exclusion from mobile networking is also easier than exclusion from territorial networking.

While the distinction between bridging and bonding is a difference in the normative power of networking, the distinction between mobility and territoriality lies in questions of proximity. Although the spatial tactic involved in exclusion from territorial networking is not that different from the one involved in exclusion from mobile networks, an uneasiness comes from the bodily proximity often involved in territorial networks: Either the deviant has to be moved, or borders have to be redrawn.

There is a difference in principle between the character of normality/deviance mechanisms in bonding and bridging networking. The more innovative character of bridging networking has to do with its weaker normative power. The strength of cross-cutting networks or weak ties (Putnam) can be said to lie in their ability to bridge diverse, normatively institutionalized social fields. Whatever the value measure, 'exit costs' are lower in bridging than in bonding networks, as bridging networkers identify less with the specific network than with the practice of networking in general. On the other hand, bridging networks are more cynical and less safe, secure or certain in terms of the meaningful reproduction of everyday life, where loose associations may be easier, but also less productive and secure, than the bonding 'marriage'. Nordic Atlantic strategies include mixes of bonding and bridging, but the Lutheran formulation of associationalism and individualism has the rationalist element of the hardworking cynical individual. While many people have been involved in religious and political revival networking with mobile bonds (there are similarities between twentieth century religious revivals and present-day NGOs), Nordic Atlantic twentieth century economic and political life has first of all been based on the territorial associationalism of local as well as national citizenship. This territorial definition of citizenship builds on a certain mix of bridging and bonding: While the individual citizen has the right to build whatever kinds of bridging networks in private and business fields he wishes, that individual is also committed to the strong, bonding local and national state.

**The sense of 'communal' identification**

Few will disagree that the formation of identities is already on the agenda in the discussion of the more or less normative character of networking. Bonding networking and bonding identification are two sides of the same coin. However, the coping-strategy approach to networking emphasizes the less normative bridging networking across institutionalized social fields. Likewise, the formation-of-identity approach focuses more on the *production* of meaning than on already institutionalized cognitive identities and identifiers. This involves a focus on society-constructing projects where innovation and the formation of identity are also often two sides of the same coin.

The socio-spatialities of coping (Figure 7.1) vary along the dimensions of innovation, networking and formation of identity. Whereas the features peculiar to innovation are often about combining mobility and territoriality, and networking is more about the social difference between bridging and bonding, the formation of identities involves the full complexity of the socio-spatiality of coping. In relation to coping, the kinds of identifications that have our interest are those that attribute meaning to coping processes involving innovation and networking. In other words, the formation of identity is about making sense of coping.

Local entrepreneurship, empowerment and culture in the Nordic Atlantic area can hardly be understood without reference to crucial local-national nexuses. The meaning of social innovation has often been linked to national identity as opposed to Danish dominance. Few cases of entrepreneurship do not have an implicit reference to nation-building. Locally, empowerment through 'managing out own affairs' has been linked to national independence.

Municipalities (*kommuner*, local/district authorities) are the most powerful regional subdivisions of Nordic Atlantic nations. Municipalities are identifications of political practices of communalism linking locality, community and place in territorial units. Rural municipalities in Norway, Iceland and the Faroes, as in Denmark, Sweden and Finland, have their historical roots in parishes. Most Icelandic, and even more so most Faroese municipalities, are still equivalent to old parishes that functioned as political, cultural and religious territorial units, even in localities with competing congregations. This is an example of how the Protestant church has been integrated in the socio-political-cultural organization of Nordic societies in a very specific way. In opposition to 'the supraterritoriality of the Roman Church' (Rokkan and Unwin, 1983:33), Protestants (Lutherans and Calvinists) merged ecclesiastical and secular administration. 'Protestantism strengthened the distinctiveness of each territorial culture by integrating the priesthood into the administrative machinery of the state within the confines of local languages' (Rokkan and Unwin, 1983:26). As a result 'communal' identification with parish municipalities has been strong in nineteenth and twentieth century Nordic peripheries, where schools, churches and local associations also have been overlapping social fields.

Territorial identifications of this kind are 'bridging' if they define people not by their membership of kinship groups, congregations or the like, but as citizens who have their own direct relations with political (as well as religious) authorities. However, the rationalist Protestant staging of reflexivity also involves the moral obligation of a 'bonding' commitment to territorially defined local and national citizenship. In these cases for those who are neither Protestants (including the secularized Protestants) nor local/national citizens, neither bonded nor territorialized, 'integration' is not easy. The relative importance of bridging and bonding varies, but Nordic Atlantic coping has certainly been a project of modernity involving territorial citizenship. This tradition is opposed to what could become twenty-first century trends of 'dealing with people' merely as professionals, taxpayers, consumers and users in regimes without territorial responsibilities (Baumann, 2000).

I have suggested that the coping strategies that have produced Nordic Atlantic societies have historical roots in the moral economies and generalized reciprocities of parish councils embedded in religious and national projects. The implicit institutionalized regulative, normative, and cognitive practices have to be seen as ways of exercising power by way of territorial boundaries at the municipal and national level. For example, Icelandic rural municipalities used to have statutory obligations to ensure that every inhabitant had work (Bærenholdt, 1991).

One may feel ambivalence about the potential essentialism that some may read out of this. The real challenge is thus whether or not this mode of socio-spatial organization is viable in an increasingly globalized world. Addressing Tönnies' *Gemeinschaft-Gesellschaft* work, Bauman raises these questions in terms of *sameness*: 'the sameness evaporates once the communication between its insiders and the world outside becomes more intense and carries more weight than the mutual exchanges of the insiders' (Bauman, 2001:13). Territorialities defined by the power of distance are 'leaking' with the development of technologies of mobility not only for bodies but – first and foremost – for information. This affects not only the small communities, but also the whole construction of a local-national nexus in Nordic societies. Some might argue that cultures and identities exist in spite of these trends. However, Bauman has a crucial point with reference to works by Fredrik Barth. Boundary drawings are not the effects of identities. The opposite is the case: '... the ostensibly shared 'communal' identities are after-effects or by-products of forever unfinished (and all the more feverish and ferocious for that reason) boundary drawing' (Bauman, 2001:17).

Mobile relations and migration in fisheries were already the rule in the nineteenth century, but as long as peripheries stayed peripheral, distances ensured territorial bonding. With the intensification of mobilities – bodily, imaginative and virtual – one cannot ensure any certainty in territorial identifications. Meanwhile, municipalities are merging to secure jobs for professionals to the 'cost-benefit' of users of public services, and new territorial and mobile orders of 'rational' administration are being introduced. In contrast to territorial bonding, people engage in mobile bonding according to profession, business, and hobby.

A number of municipalities in Iceland, the Faroes and coastal Norway have received immigrants because of the shortage of labour power in low-status fish-processing industries, and in this way some of the most 'peripheral' localities are being internationalized. In other cases, foreigners play entrepreneurial roles in the development of tourism thanks to their international networks and language abilities. In both kinds of cases, though, it is because 'outsiders' take roles left vacant by 'insiders' that they are accepted, and they do not gain citizenship rights through this alone. In some Icelandic cases, multiculturalism is celebrated in the periphery (Skaptadóttir, 2004), whereas in some North Norwegian cases, immigrants from

Russia have for example been excluded from swimming facilities because of the health risks they are said to carry with their bodies, while Russian women are generally presumed to be prostitutes (Aure, 2004).

Clearly, 'communal' identifications are being challenged, and it is not obvious how mobilities are supposed to be bridged or bonded territorially. To further examine and illustrate these points, the following sections are two short accounts of the networking involved in cases of innovation in Greenland and Iceland, where identifications do not fit the socio-spatial practices involved in innovation and networking, and thus challenge the ideal of coping.

### The disorders of mobile success in Ilulissat, Greenland

The socio-spatial practices involved in the commercial development that is supposed to contribute to the progress of a Greenlandic society striving for autonomy create significant cases of displacement of the territorial organizations that took the initiative for such developments. In the Greenlandic town of Ilulissat (the third-largest in Greenland with just over 4000 inhabitants), the municipal council and mayor took the initiative to create a new business, 'Arctic Fish'. At the end of the nineties this firm, originally 'midwifed' by municipal authorities, became one of the prime movers in Greenlandic crab fisheries for international markets. The director is a headhunted, experienced and committed Danish businessman, local fishermen are shareholders, and factory ships only employ workers from the municipality. Incomes are very high, and the ships operate in waters distant from Ilulissat (other municipalities are in fact closer to these waters). This was the situation in the spring of 2001, when the entrepreneurial Social Democrat ('Siumut') mayor lost local elections to a populist grouping outside the established political parties (Bærenholdt, 2002a). The 90 local citizens who work on the distant factory ship earn high incomes, which also generate high income taxes for the municipal authority, which has now recovered its original investment, for a feasibility study, more than a hundredfold.

Still, other local citizens are not 'integrated' in the local labour market. Social problems, suicide, violence, lack of family care for children etc. are among the issues with which the Home Rule Government and its Minister of Social Affairs have to cope in many localities including Ilulissat. The National Association of Municipalities has asked for the creation of more space in institutions for children and youngsters who have to be placed in care by the authorities, and the Minister has asked the municipalities to report on their efforts with local policies for children and youngsters (as reported in a Nordic magazine, Schultz-Lorentzen, 2002). Not everyone gets a slice of the pie, but even for those who do, this is not always meaningful.

The very same locality is the major Greenlandic tourist destination. Ilulissat is located beside the most productive 'ice fjord' in the Northern Hemisphere. This is where the movie *Smilla's Sense of Snow* was shot, and the *Smilla* name is used for several tourist facilities, including the facilities left behind by the camera crews. These have been supported by local outfitters (Greenlanders and Danes) who also offer tours to tourists. As in many other places, there are major problems of cooperation in local tourism, and part of the picture is the rivalry between the foreign (non-Danish) entrepreneurs and the Home Rule Government-supported Danish-Greenlandic tourism businesses (Bærenholdt, 2002a). These small stories from the municipality of Ilullisat are illustrative, although no claims have been made about causalities. Arctic Fish and local tourism development were investigated in a comparative Nordic project (Bærenholdt, 2002a and 2002b). Ilulissat is not one community. The business milieu has been developed as a result of municipal entrepreneurship, but it is not embedded among local people. While *territorial bonding* among local fishermen, businesses and some workers has been as important as *mobile bridging* with international business partners, the gap between losers and winners raises the question of the need for *territorial bridging*. With the kind of entrepreneurialism and global orientation that innovative municipalities are pursuing in Ilulissat, as in many other places in the world, the normative power of 'people's' strategies (and its academic conceptualizations such as 'empowerment' and 'embeddedness') becomes ambivalent. Disembedding is what is happening, and if the mobile bonds of multinational firms are introduced, the socio-spatial order of Nordic Atlantic municipalities will be a thing of the past.

### The role of territoriality in Iceland's mobile order in Hornafjördur

In the socio-spatial production of Iceland, which is truly a product of nationalist strategies, the 'closing' of the circle of the national Ring Road in 1974 was a symbolic event. The event took the form of a physical, social and symbolic *bridging* of (and in) the (now former) rural municipality of Öræfi with the west via bridges over the many glacial rivers over Skeidararsandur just south of the Vatnajökull, 'Europe's largest glacier'. The bridges were constructed so that future floods from the subglacial volcanic eruptions (this happens every decade) would open and disconnect but not totally destroy them.

Many Icelanders are internationally well known for their national identity, their mobility and their work ethos. Mobility by means of planes and automobiles connects the various parts, and expresses the territorial control, of Iceland, while Icelanders take their national identity with them when they travel the world. In simple terms, I have an impression of Iceland as a corporate national business bonding of global citizens (Bærenholdt, 1991).

The Icelandic municipality of Hornafjördur (with over 2300 inhabitants) 'covers a 200 km long, but narrow, strip of land on the southeastern side of the ice cap of Vatnajökull' (Benediktsson and Skaptadóttir, 2002). There used to be six municipalities: the town of Höfn (1800 inhabitants) and five small rural (parish) municipalities. They merged during the 1990s, and this also meant that the municipality took over more functions from the state. The amalgamation meant that the scattered farms of the former Öræfi municipality and of the twentieth century fishing town of Höfn came together. Two cases of local development, one in Öræfi and one in Höfn (and other places), illustrate the role of territoriality in Iceland's mobile order along the Ring Road (Benediktsson and Skaptadóttir, 2002; Jóhannesson, Skaptadóttir and Benediktsson, 2003).

Agriculture has been through major restructurings in Iceland since 1979, and as in many other areas, tourism has become a major alternative for making a living in rural areas. The completion of the Ring Road turned Öræfi into a different place, now much easier to reach by car for touring and passing Icelanders and international tourists. In Öræfi, the Skaftafell National Park is the major tourist attraction, placed as it is just beneath the Vatnajökull glacier. A number of local people and their families have engaged in the construction of rural hotels, the development of special 'coast-to-mountains' tours, and the introduction of a local waste incinerator. The incinerator burns the waste from visiting tourists (as well as locals) and the energy from this is used to heat a swimming pool that is attractive to tourists (as well as to locals). There are a number of small-scale local businesses that cooperate through local territorial networking. The inclusion of Öræfi in the amalgamated municipal district of Hornafjördur has not been popular, as entrepreneurs think the distance from the municipal centre is too great, not only in kilometres but also culturally between the (old) rural area and the (young) fishing town. Local tourism development depends totally on the traffic on the Ring Road; mobilities outside its control nourish the local economy in a contingent way. Special events such as the 1996 subglacial volcano eruption and the subsequent flood brought extra visitors to gaze at the event. Mobile bridging is only an issue of participation in tourist marketing, while the driving forces are the attraction of the landscape and the access made possible by the Ring Road. In this case we are talking about roadside social-spatial practices of territorial bonding (of locals) *and* mobile bridging (with passing tourists), drawing on the territorial bridging of infrastructure development.

Another project has been the biotechnological venture 'NorthIce', which was established thanks to the conditions produced by the municipal authorities, the local co-operative society and others in cooperation with the University of Iceland in Reykjavík. Apart from the raw materials needed to produce enzymes and seafood flavouring ('NorthTaste') for the catering market, there were few reasons to locate the business in Höfn. But there were also few reasons not to do so. NorthIce is a case of business location facilitated by personal networks of a non-business character. This is mobile bonding practice, although produced by a bridging between different milieux. It also means that production can be stopped or removed. So far, the free flow of information between authorities and businesses has facilitated development, but the effects are uncertain (Benediktsson and Skaptadóttír, 2002). The territoriality involved has been created by the efforts by municipal authorities as well as pushes in Reykjavik, in addition to the territoriality involved in the personal social networks.

In the Icelandic cases, we see dynamics of contingent developments facilitated by the mobility and territorial bridging involved in the use of the Ring Road. As long as Öræfi tourism development combines strategies of mobile bridging with markets and the territorial bonding of local people, it is pursuing the traditional modern strategy of Nordic Atlantic local development based on locally-controlled natural and financial resources combined with flows into international markets. This is also the case with tourism development in Ilulissat. Hence, the cases of Arctic Fish based in Ilulissat and especially of NorthIce in Höfn involve stronger mobile business bonds, and this means that the degree of territorial commitment in these ventures is limited. The role of municipal authorities in innovation is limited to facilitating or 'midwifing', as it is called in Greenlandic. However, the settlements are totally dependent on social, educational and other services maintained by municipal authorities.

## Coping beyond the modern Nordic periphery?

> The concept of society will in the future be one particularly deployed by the especially
> powerful 'national' forces seeking to moderate, control and regulate their variously
> powerful networks and flows criss-crossing their porous borders (Urry, 2000:1).

What are the socio-spatial practices in the coping strategies exemplified in the projects mentioned? The regulation effort in the Nordic periphery was never the task of nation-states alone, since the 'local states' (municipalities) played a significant role in the process of modernization. In fact, the borders are very porous to the powerful networks and flows of tourists and fish companies, and they are so because coping strategies include networking and innovation among these actors. The cases show combinations of territorial bonding and mobile bridging, but the bonds constructed are only 'parts of' municipal territorial bonds. The bonding involved in Arctic Fish and other ventures uses the municipal identities of their projects, but in so doing the actors only *talk about* the inclusion of those who are in fact excluded. Of course the municipal tax from incomes earned in Arctic Fish can be used for social services, and perhaps the displacement from their families of children and the young. This is the redistribution of the welfare state, and as long as it works, the concept of society as a concept of regulation, rights, duties and redistribution among citizens has a meaning.

As such, coping in modern, Nordic peripheries has relied on societal relations that adhere to principles of association (Aarsæther and Bærenholdt, 2001; cf. Mingione, 1991). Indeed, associational relations organized according to territorial principles have been used to regulate and moderate the networks and flows from which people have earned a living. The territoriality involved in this regulation is *not* that of bonding identifications of (for example nationalist) 'sameness' – but that of territorial bridging citizenship of Nordic territorial states.

Municipalities do play both innovative and re-distributive roles in the Nordic peripheries, but they only do this by means of redistribution via state and municipal taxes. Identification with Protestant ethics, parish communities, and their very status as peripheries have been elements in their historical socio-spatial production. Now these elements have become threats to those same societies, because they would fence off Nordic peripheries from the forces of innovation, networking and identification, from bridging and mobile coping strategies. Still, redistribution across spatial and social difference and uneven development produces societies through the territorial boundaries of such regulation. What these regimes have managed to do is both to include and exclude the poor through principles of territoriality, while they have also mobilized the rich.

In response to Laxness' (Gudmundsson's) motto we can now observe that '*no difference is greater than the one that separates citizens within the same munici- pality*'. Distance, disease and poverty have been killed off; the result is a territorial modern order of 'holding people together', which only makes the differences in mobility between them all the more peculiar. Indeed, concepts of empowerment and embeddedness have become ambivalent. To stress the differences (as opposed to the distances) within each municipality (as opposed to each country), is to stress the uncertain but real possibility that municipal 'transformers' can cope with mobile social relations by applying principles of territorial bridging between those who are empowered/embedded and those who are not. Neither territorial 'communal' identifications nor mobile bonds, businesses, and brands make any sense to people if they do not signify coping processes that sustain their livelihoods. Since there is no 'one people', there is no singular 'empowering' and 'embedded' coping strategy for the inhabitants to follow.

Peoples of the Nordic peripheries coped so well with distance in the twentieth century by applying principles of territorial bonding, but this way of coping is now of less use. Since mobile bridging coping is usually for the few, the challenge of coping in twenty-first century 'post-peripheries' seems to be that of combining the mobile bonding of businesses and the territorial bridging of municipalities in meaningful ways.

## Acknowledgement

Thanks to Kirsten Simonsen and Keld Buciek for their constructive comments on a number of earlier versions. All remaining shortcomings are my responsibility.

## Notes

1   See the websites www.uit.no/mostccpp and www.unesco.org/most/p91.htm.
2   Here I draw on 2001 case studies of projects of local development in Iceland, Greenland and the Faroes from the NORDREGIO (Nordic Centre of Spatial Development) project 'Coping Strategies and Regional Policies, Social Capital in Nordic Peripheries' (Benediktsson and Skaptadóttir, 2002; Bærenholdt, 2002a and 2002b; Hovgaard, 2002; Jóhannesson, Skaptadóttir and Benediktsson, 2003). Parts of these case studies are presented in more detail in later sections.

## References

Aarsæther, Nils and Bærenholdt, Jørgen Ole (eds), 2001: *The Reflexive North*, Copenhagen: MOST and Nordic Council of Ministers, Nord 2001:10.
Aure, Marit, 2004: 'The Transnational North: Constructions of Labour Migrants', paper from PhD project submitted to journal.
Bauman, Zygmunt, 2000: *Liquid Modernity*, Cambridge: Polity.
Bauman, Zygmunt, 2001: *Community, Seeking Safety in an Insecure World*, Oxford: Polity.
Benediktsson, Karl and Skaptadóttir, Unnur Dís, 2002: *Coping Strategies and Regional Policies – Social Capital in the Nordic Peripheries – Iceland*, Stockholm: NORDREGIO Working paper 2002:5.
Braudel, Fernand, 1981. *The Structures of Everyday Life – The Limits of the Possible*, Civilization and capitalism 15th–18th century vol. 1, New York: Harper and Row.
Bærenholdt, Jørgen Ole, 1991: *Bygdeliv*, Roskilde: Institut for Geografi, Samfundsanalyse og Datalogi, Roskilde Universitetscenter, Forskningsrapport nr. 78.
Bærenholdt, Jørgen Ole, 2001: 'Territorialitet, mobilitet og mestringsstrategier', in Kirsten Simonsen (ed.): *Praksis, rum og mobilitet – socialgeografiske bidrag*, Copenhagen: Roskilde Universitetsforlag, pp. 141–171.

Bærenholdt, Jørgen Ole, 2002a: *Coping Strategies and Regional Policies – Social Capital in the Nordic Peripheries – Greenland*, Stockholm: NORDREGIO Working paper 2002:4.
Bærenholdt, Jørgen Ole, 2002b: *Coping Strategies and Regional Policies – Social Capital in the Nordic Peripheries*, Stockholm: NORDREGIO Report 2002:4.
Bærenholdt, Jørgen Ole and Aarsæther, Nils (eds), 2001: *Transforming the local, Coping Strategies and Regional Policies*, Copenhagen: MOST and Nordic Council of Ministers, Nord 2001:25.
Bærenholdt, Jørgen Ole and Aarsæther, Nils, 2002: 'Coping Strategies, Social Capital and Space', *European Urban and Regional Studies*, 9 (2):151–165.
Bærenholdt, Jørgen Ole and Haldrup, Michael, 2003: 'Economy-Culture Relations and the Geographies of Regional Development', in Öhman, J. and Simonsen, K. (eds): *Voices from the North*, Aldershot: Ashgate, pp. 49–66.
Cloke, Paul and Jo Little (eds), 1997: *Contested Countryside Cultures*, London: Routledge.
Cooke, Philip and Morgan, Kevin, 1998: *The Associational Economy: Firms, Region, and Innovation*, Oxford: Oxford University Press.
De Certeau, Michel, 1984: *The Practice of Everyday Life*, Berkeley, LA/London: University of California Press.
Hovgaard, Gestur, 2002: *Local Coping Strategies and Regional Policies – Social Capital in the Nordic Peripheries – Faroe Islands*, Stockholm: NORDREGIO Working paper 2002:8.
Jessop, Bob, 2000: *Globalisering og interaktiv styring*, Frederiksberg: Roskilde Universitetsforlag.
Jóhannesson, Gunnar Thór; Skaptadóttir, Unnur Dís and Benediktsson, Karl, 2003: 'Coping with Social Capital? The Cultural Economy of Tourism in the North', *Sociologia Ruralis*, 43 (1):3–16.
Mingione, Enzo, 1991: *Fragmented Societies, A Sociology of Economic Life beyond the Market Paradigm*, London: Basil Blackwell.
Paasi, Anssi, 1996: *Territories, Boundaries and Consciousness*, Chichester: Wiley.
Putnam, Robert D., 2000: *Bowling Alone*, New York: Simon and Schuster.
Rokkan, Stein and Unwin, Derek W., 1983: *Economy, Territory, Identity – Politics of West European Peripheries*, London: Sage.
Sack, Robert D., 1986: *Human Territoriality: Its Theory and History*, Cambridge: Cambridge University Press.
Schultz-Lorentzen, Christian, 2002: 'Massivt omsorgssvigt' [Massive lack of care], *Politik i Norden no. 2*, magazine published by the Nordic Council and the Nordic Council of Ministers, p. 25.
Skaptadóttir, Unnur Dís, 2004: 'Mobilities and Cultural Difference: Immigrants' Experience in Iceland', Ingimundarson, V., Loftsdóttir, K. and Erlingsdóttir, I. (eds) *Topographies of Globalization: Politics, Culture, Language*, Reykjavík: University of Iceland Press.
Stenius, Henrik, 1997: 'The Good Life is a Life of Conformity: The Impact of the Lutheran Tradition on Nordic Political Culture', in Sørensen, Øystein and Stråth, Bo (eds) *The Cultural Construction of Norden*, Oslo: Scandinavian University Press, pp. 161–171.
Storey, David, 2001: *Territory, The Claiming of Space*, Harlow: Prentice Hall.
Therborn, Göran, 1999: 'Modernity, Globalization and the Rural World', in Charalambos Kasimis and Apostolos G. Papadopoulos (eds): *Local Responses to Global Integration*, Aldershot: Ashgate, pp. 21–39.
Urry, John, 2000: *Sociology Beyond Societies, Mobilities for the Twenty-first Century*, London/ New York: Routledge.
Virilio, Paul, 1998. *Cyberworld – det værstes politik*, Copenhagen: Introite!

Chapter 8

# Mobility and Territorialising Regimes in the Andes

Fiona Wilson

My aim in this chapter is to discuss how space is constituted and produced at the margins of postcolonial states by taking an example from Latin America. This involves investigating how different social configurations produce qualitatively different conceptions of space at two scales of analysis: at the level of rural/peasant society and at the level of the uneasy zone of contact between peasants and the overarching territorialising regimes of which they form a part – and from which they often wish to distance themselves or escape. This line of thought has been prompted by reflections on longer-term fieldwork carried out in the central Peruvian Andes with groups of peasants who became displaced on account of civil war in the 1980s and 1990s. The rural groups had lived in relatively inaccessible localities, linked to the outside world mostly by tracks and trails, only occasionally by dirt roads. In the civil war, they suffered two invasions. The first was by Maoist insurgents who attempted to reconfigure the rural areas as their support base and bring them into an oppositional or mimic state; the second was by the state's military forces whose objective was to stamp out subversion and bring rebellious Andean provinces back into a state of government. From a position of relative autonomy on the margins of the state, these rural societies were incorporated violently into the territory of the state. What can be discerned about these processes of change at the level of both state and peasantry by focusing on the production and destruction of social space?

## Territorialising regimes: state versus the peasants?

There are two conceptual questions guiding this discussion posed at different analytical levels. The first is what happens to understandings of social life when movement and mobility are put at the centre? This has long been a preoccupation of geographers as well as some sociologists and anthropologists, yet still represents an unorthodox approach in inter-disciplinary development studies. We can start by raising questions about the nature of connections. How is movement channelled? What kinds of routes or roads emerge with respect to different kinds of political economy? A distinction needs to be drawn between tracks and trails on the one hand, and paved roads that can be transited by wheeled and motor vehicles on the other. While tracks and trails are institutionalised yet flexible demarcations of flow, roads are more permanent features of the landscape that fix (and hopefully) facilitate lines of flow. The construction and maintenance of roads require much higher levels of organisation and investment, command over labour and technical know-how than tracks/trails. They also possess much greater symbolic and material value for states – often becoming a vital element in the state's territorialising project of extending sovereignty up to the frontiers.

Secondly, one needs to look at movement not only with respect to routes but also as spatial practice related to enveloping, encompassing configurations and organisations of space. Without such structured reference points (fetched up against and/or negotiated by social actors), analysis of movement easily becomes a meaningless exercise. People think and move within definitions of space that are by no means natural or universal, but are underpinned and supported, structured and contained by concepts and practices of spatial politics with their own historical trajectories (Agnew and Corbridge, 1995). To describe such arrangements and organisations of space, I propose using the concept of territorialising regime, a term that suggests the production of social space (Lefebvre, 1991); political fields constituted by the interplay between dominating forces and oppositional struggles (Bourdieu, 1991); and geo-political re-ordering that takes place in and reworks different spatial scales (N. Smith, 1992). Territorialising regime is thus an attempt to capture the materialized, spatialised characteristics of a particular kind of political economy.

A territorialising regime can be defined as a non-scale specific, spatial/political construct. As a *regime*, some form of centralised political organisation is assumed to exist within a political field whose hegemonic authority rests on shared understandings as to the ideas, rules, practices and contests structuring the political game. This certainly does not imply harmony or consent, for the notion of hegemony can be extended to incorporate contested meanings and struggle: 'the ways in which the words, images, symbols, forms, organisations, institutions and movements used by subordinate populations to talk about, understand, confront, accommodate themselves to, or resist their domination are shaped by the process of domination itself' (Roseberry, 1994:360–1).

As a *territorialising* entity, not only is land/territory taken as root and symbol of belonging, it becomes imagined and materialised as a dimension over which political control can be exercised (however 'uprooted' actual society happens to be). Territorialising is an endless process, a process without closure, in which claims to political/social space are constantly being brought into play (and questioned) through colonisation, maintaining of routes, tying of people even when living at a distance. Here one must allow for a measure of slippage between the envisioned and the concretised; between imagined space as configured in the mind's eye of the powerful and in the maps of the planners, and the actual messy, untidy, conflict ridden, power-laden landscapes inhabited by the 'territorialised'. Analysis of territorialising regimes involves a double move: enquiry into how spatial organisation and practice structure power relations and how power relations underlie, mould, direct and reverberate in space. In this, as Allen (1999) argues, strategies of force as well as persuasion and seduction come into play when the powerful try to impose/demarcate boundaries, classify regions and populations, and guide, deploy and fix people in space.

We are so used to the dominating spatial ordering of the modern state that it is almost taken for granted. In line with their territorialising projects, states claim sovereignty, establish political economies, provide founding/unifying myths of community, and cover the bounded domain with hierarchic administrative divisions and agents/representatives to rule, govern, administer and defend. A dominant territorialising regime (like the state) seeks to create order at one scale (the national) both by becoming a translocal presence and by implanting localised offices, institutions and practices in which the state is instantiated (Gupta, 1995:375–6). States define territory in terms of self-evident sovereignty whose varied component elements (regions, cities, natural resources) are constructed into an integrated whole (Radcliffe, 2001:123). But seen in spatial terms, state sovereignty is only 'quasi sovereignty' at best (Allen, 1999:197).

One advantage of thinking in terms of territorialising regimes is that it allows for a consideration of geographical scale and conception of superimposed or interpenetrating configurations and images of space. Instead of working with fixed, static classifications like the familiar micro, meso, macro boxes we can suggest, following Neil Smith (1992:72), that 'social life operates in and constructs some sort of nested hierarchical space rather than a mosaic'. As a territorial regime, the state is under constant reconstruction; it advances and retreats, becomes reconfigured on the margins, achieves a presence in some places but never succeeds in capturing others (Hansen and Stepputat, 2001). Over-arching regimes like the state can inspire attempts by the dominated to liberate themselves, and they do this by harnessing powers and instrumentalities at other scales. In relation to recent mobilisations, protests and movements emerging at local and regional levels in Andean countries, Slater (1998) argues that new associations are being made between democratisation and decentralisation and in the struggle against centralism new forms of spatial subjectivity have emerged. 'These new forms, which contest the given territoriality of the political system, can be viewed as reflections of political expressed spatially', in response central state administrations have been forced to introduce reforms seeking to contain and incorporate these local and regional resistances (Slater, 1998:387).

Although propositions as to the formation and character of territorialising regimes can be presented in relation to state sovereignty, they also underlie an old discussion as to the specific nature of territoriality and social space in Andean society. The pioneering anthropologist, John Murra, from the 1950s delineated what (in effect) was the territorialising regime of Andean ethnic polities prior to Spanish Conquest and this later became known as the model of vertical control.[1] Murra emphasised how the economic and political systems of ethnic polities in pre-colonial times were based on taking direct control over a maximum number of ecological levels. Resource heartlands were mostly located at around 3,000 metres where maize and potato lands met. Since this was a social formation where the market did not exist, management of the vertical economy demanded the deployment of individuals and households who belonged to the principal community by a central authority (the chiefs) to colonise and settle in archipelagos of disparate resource islands, located at a distance, above and below the central settlement. In these islands, resources were shared amongst several ethnic groups. The physical boundaries of ethnic domains remained fluid and porous, but mobility and deployment of people under this Andean territorialising regime were strictly controlled.

While Murra worked in the archives, his students in the 1970s set out to investigate the survival of the model of vertical control within contemporary Andean society. Many argued that this spatial/political organisation had perpetuated a specifically

Andean way of doing things, and that cultural continuity was more salient than change. The way of managing resources in a mountainous landscape was considered an enduring cultural feature notwithstanding the violent imposition of higher order territorialising regimes by a series of invaders: Inca, Spanish and post-colonial governing elites. Studies conducted at various scales found evidence of verticality: at the levels of the household, kin-based collectivity, land-based community (the latter a product of Spanish colonial rule). This did not signify tradition, rather that indigenous society had sought continuity as a tactical way of opposing the colonising projects that threatened to engulf them.[2] In an essay published in 1982, Bradby offered a timely clarification of much of this literature. When discussing Andean social organisation through the lens of verticality, she argues that the crucial issue is not the banal one of how 'vertical' resources in mountainous terrains are accessed, but the much more intriguing and problematic issue of political authority and control. How under different over-arching territorialising regimes was local political authority exercised and maintained, surplus appropriated, and political, ideological and religious power created and preserved?[3]

   This chapter intends building on this old debate as to the constitution of space and nature of political control over space in the Andean context. I shall focus in particular on one aspect that was never clearly worked out in the earlier literature: how mobility and control over movement (as exemplified in labour deployment) produces space and underpins spatial relations and imaginations. The chapter will address two questions: (i) at what scales and through what forms of control and authority have territorializing regimes in the Andes conditioned mobility; and (ii) how have rural populations confronted overarching/overlapping regimes that sought to territorialise them? I shall discuss first the over-arching territorialising regimes characteristic of Andean political economy since colonial times. Then, taking the example of Cayash, a cluster of land-holding communities on the northern perimeter of the province of Tarma, Peruvian central Andes, I shall explore how rural populations fashioned their own conceptions, perceptions and usages of space and how these practices were interwoven with and at the same time were opposed to the over-arching territorialising regimes in which these populations were placed.

## A short history of mobility and territory in the Peruvian Andes

Over the plurality of ethnic polities that managed vertical economies on roughly similar lines, the Incas built an empire. A central feature of the new imperial regime was the right assumed by Inca governors to deploy subjects over an extensive territory. Ethnic groups were sent far from their homelands to work important resources demanded by the Inca, to populate the frontiers after military conquest and to diffuse opposition. Spanish colonial rule built on this foundation, but was much more destructive. Colonisation led to catastrophic population decline on account of European disease and the brutal treatment of indigenous people (that reduced the population of what became later the territory of Peru from some 10 million to 1 million inhabitants in the period 1530–1600). The relative autonomy of the vertical economies was broken down and the ethnic groups were brought under the command of Spanish ecclesiastical and military orders. Some were re-designated as Indian parishes and communities, placed under the protection of the Spanish Crown and compelled to provide tribute and labour service, others were handed over as grants of land and people (haciendas) to Spanish settlers. Spanish officials re-modelled and expanded the deployment of indigenous subjects in colonial space by instituting a massive system of forced labour. Indians were to service the needs of the Spanish towns and provide the manpower to work the mining and other enterprises that were the mainstay of the Spanish colonial economy. By law, tribute-paying Indians had to leave their communities to work one year in seven in the silver and mercury mines; as the native population succumbed to disease, so forced labour became increasingly burdensome on the survivors.

Strategies of labour deployment and territorialisation implanted by the Spanish had dramatic, unintended consequences for the response of Andean peoples caught in the blast of the Spanish colonising project was to move. Physical movement became a way to achieve fiscal reclassification. From being hard-pressed *originarios*, or founding members of an Indian community, compelled to provide tribute and labour service in return for usufructuary rights to land, families sought re-classification as *forasteros* (newcomers initially deprived of land) or as hacienda residents (bound through personal service to a Spanish lord), both classes being exempt from tributary obligations to the colonial state. Others sought reclassification as mestizos, literally half castes, who moved into town, took an urban artisan trade and changed their ways of dress and consumption.

Processes engendered by colonial rule produced complex reactions, appropriations and adaptations on the part of indigenous populations. One of these followed from the introduction of the market. As Larson (1995:10) writes: 'Out of the imperatives of the silver economy and the pre-existing capacity of Andean ethnic groups to produce surplus goods and services, markets took root and spread throughout the Andes, where money and mercantilism had never before existed.' But market relations never wholly displaced the relations of exchange built up prior to the conquest. Another adaptation followed from the institutionalisation of property rights that under the Spanish were fixed and given the force of law. This produced a new way of conceiving space as landed property registered in legal documents, to which the Crown as symbolic head awarded and protected rights. These moves reflected not so much an imposition by a scatter of colonial officials, but a kind of settlement, a *modus vivendi*, reached by the indigenous aristocracy (still allowed many privileges up to the end of colonial rule) and Spanish authority. Indigenous peoples turned to the law and the law court, and made them key sites for working out and holding onto their own territorial claims that pitted the indigenous land-holding community against the hacienda. But the imagination of space as property never displaced the older conception of space as formed by communal rights and safeguarded by indigenous forms of control over movement. Since Spanish rule depended on political alliances with the indigenous aristocracy, this offered some protection and possibility of survival. Although some ethnic groups and kin-based clans disintegrated, others struggled hard despite the harsh new circumstances to re-order their social and spatial organisation and avoid a complete rupture with the past.

Three main points concerning mobility can be drawn from this colonial history. First, although the colonial territorialising regime was imposed by force, indigenous territorialising regimes endured at a different geographical scale. Overarching regimes were never coherent, fully realisable spatial projects. Past regimes continued to leave traces, detritus, in the landscape and this facilitated the survival of a memory as to alternatives. Different regimes could co-exist precisely because they occupied different political, social and symbolic registers. Regimes became particularly clear – and particularly ambiguous – on the margins, in arenas of encounter between the deployment strategies of the powerful and the subversive tactics of subjects.

Second, territorialising regimes could be distinguished not so much by their physical boundaries but through their conditions and relations of mobility. Under the Spanish colony, indigenous movement took place primarily in three spatio-political fields. The first was through 'official' strategies of labour deployment where indigenous people were forced to move over great distances to work on state projects and/or earn tribute. Control over this movement was foundational for the territorialising regime of the colonial state. The second was through the unintended consequences of forced labour when Andean peoples resorted to movement as a tactics of resistance to evade and dilute the powers of the dominating regime. While this mobility in part undermined the colonial state, it led people to seek refuge on the haciendas, where they were exempt from paying tribute and labour service to the state. The third was the relations of movement characteristic of the indigenous vertical economy on the margins. In none of these situations was movement free or voluntary.

Third, movement had become imbued with racist classifications that changed over the colonial period. By the time of the Wars of Independence in the early nineteenth century, mestizos and Indians were seen by the white elite to occupy opposite poles in terms of propensity and capacity for autonomous movement. Mestizos, though despised and marginalised in colonial discourse, were increasingly accredited as transporters (muleteers), traders and brokers, moving at ease between town and countryside (Harris, 1995). An opposing view of the indigenous population had gained currency. Partly in an attempt to erase memories of the massive indigenous insurrections against the colonial state in the late eighteenth century, a new image of the Indian was conjured up. The stereotypical Indian was depicted as either enclosed in 'his' community or hacienda, or as a rootless (uncivilised) wanderer alone with his llama. By the late nineteenth century, 'scientific' racist ideology attributed Indian-ness increasingly to the remote, uncivilised 'others', who were insufficiently autonomous in action or thought to be worthy of citizenship. This representation of the postcolonial Indian as impoverished, uncivilised and immobilised became entrenched within national and regional elites who strove to re-instate coercive forms of labour deployment at the level of the state. Forced labour was considered necessary for the nation's mines, plantations and to build grandiose infrastructure projects that aimed to bring modernity and progress by driving railways and roads from the Pacific coast, over the Andes and down to the upper Amazon basin.

After Independence, white elites and foreign enterprises staged new onslaughts on indigenous resources now that they were freed from protection by the Spanish Crown and supposed to be incorporated into a national capitalist market. Some indigenous-peasant populations lost lands and livelihoods as a result of the expansion of merchant capitalism, territorial aggrandisement of haciendas and perpetuation of tribute and forced labour (under new names and guises). But in some regions, especially in the central Andes, revival of the mining industry and rapid expansion of a plantation economy in the Amazon lowlands (producing cane alcohol and coffee) opened up many new income-earning opportunities that were eagerly grasped by nearby indigenous communities. New trading patterns and relations meant that regional and national markets became re-configured, a process favouring some social groups and some routes and transport nodes while relegating others to decline and oblivion. The concept of accessibility became linked to modernity. But at the same time, the postcolonial state had lost all interest in bringing progress – later re-labelled development – to the Andean hinterland. This was a territorialising regime directed by the capitalist market and foreign enterprise, a regime in which the state was not only virtually absent but also quite ineffective in legislating or regulating the rules of the game.

The dualistic socio-spatial organisation characteristic of the colonial period, rooted in the distinction between haciendas and communities, remained. Haciendas and their owners remained power points in the landscape. Only when regional oligarchies began abandoning their Andean properties, preferring to invest in more lucrative ventures elsewhere, did the hacienda system begin to crumble and implode. The 1960s saw some increasing state interest in the Andes, but it was too little, too late. The decade was marked by strident demands for agrarian reform and the nationalisation of foreign enterprise. Simmering unrest and active guerrilla *focos* prompted the army to take state power through a coup in 1968. A reformist, nationalist military government was installed and with the assistance of radical, left-wing advisers nationalised foreign firms and implemented sweeping agrarian and educational reforms. There was widespread belief that a mixture of structural change and state-led popular participation could propel socio-economic progress and remove 'feudalism' from the Andes. Suddenly the state was far more highly visible in the Andean provinces than ever before, embodied by the officials and advisers who toured the countryside in their Land Rovers. But political response was muted as dictatorial measures curtailed all practice of local democracy.

Agrarian reform though solving some of the peasantry's problems generated many new ones, and this created amongst some a greater readiness to take violent action (Kay, 2001). After the fall of the reformist government, there was neither interest nor resources on the part of the state to assist in the reconstruction of Andean society. After the land-owning oligarchy retired to the coast, they left a hiatus, a vacuum, with respect to the overarching territorialising regime. This 'empty' space, some say, was then open for colonisation by revolutionary political projects, amongst which those of the Maoists became especially prominent. But the attribution of hiatus did not necessarily make much sense for peasant communities who continued organising space and imagining a state of their own choosing.

What can be concluded about the nature of colonial and postcolonial territorialising regimes? One point to stress is that they had brought ruling and subject populations into the same meaning universe with regard to respect for the law, concept of territory as property and market relations. However, the postcolonial regimes had failed to instil any deeper sense of nationalism, or nationalism as an expression of cultural intimacy (Herzfeld, 1997) capable of over-riding the social gulf distinguished by attributions of 'race'. Authoritarian power structures remained in force at different scales and social levels and with this, command over resources as well as over movement perpetuated segregated forms of spatial practice. Though some attempt at re-labelling was made under the reformist military regime, when Indians were renamed peasants, attributed racial difference continued to underlie, direct and divide the production of social space.

## The peasantry and oppositional space

In the second part of the chapter concepts of territorialising regime, production of space and conditions of mobility are used to frame a discussion of how a particular group of indigenous peasants organised their lives and livelihoods and confronted the outside world. The text is built up from accounts told by people from Cayash, hamlets/communities whose population numbered some 600 families immediately prior to civil war. I shall reflect on conversations with families who had left Cayash in search of refuge in the provincial capital, Tarma town, after the military invasion of 1991. The period discussed is the so-called hiatus, spanning the end of hacienda rule, years of community autonomy, invasion by *Sendero Luminoso*, violent military reprisals, displacement to the towns, and current phase of limited resettlement. Three themes will be addressed. First, from a *Cayashino* perspective how are past territorialising regimes perceived and what kinds and conditions of mobility are associated with them? Second, what light is thrown on the nature of social space when we re-trace the journeys and analyse accounts of spatial practice from two generations of *Cayashinos*? Third, can the armed conflict be seen as inaugurating the onset of a new state-led territorialising regime and what place might *Cayashinos* have in it?

## *Cayashino* encounters with territorialising regimes

The six hamlets of Cayash dispersed in valleys of the Ulcumayo river system are located at roughly 3,200 metres, above the precipitous drop down to the humid lowlands. The communities were in possession of more that 30,000 hectares and each could count on ample extensions of prime arable land for growing many varieties of potatoes, the coveted *papas de color*, as well as other Andean tubers (*olluco*, *mashua*, *oca* and *maca*) and pastureland. Households were allocated communal land annually by the authorities in accordance with a six-year crop rotation (where the land is planted with potatoes for two years and left fallow for four), as well as own private plots. The *comuneros* possess a (now) rare knowledge of how to manage potato production without the application of chemical fertilizers or pesticides. The agrarian system of Cayash is very different from that of neighbouring Huasahuasi, long the most important commercial producer of seed potatoes in Peru, that has faced intractable problems due to contamination of the soil. One explanation of the perpetuation of product diversity and sustainable resource management can be linked to the region's relative inaccessibility and autonomy.

No roads connect the hamlets to the world outside; *Cayashinos* must walk some 50–60 kms south to reach the tarmac road to Tarma, or 25–30 kms north to reach a dirt road leading up the Ulcumayo valley. All goods moving in and out are transported by pack animal. An attempt had been made by the modernizing central state to build a road down the Ulcumayo valley in the 1890s, with the aim of controlling – i.e. taxing – the movement of cane alcohol sent from the lowland plantations to the highland mining camps. But communities in the region resented being dragooned into forced labour, engaged the prefect's troops in bloody battles and stopped the road-building scheme. A few roads were later built by foreign mining companies but only in the 1960s did the central state give limited support to the construction of feeder roads; these, however, stopped well before the borders with Cayash.

The older generation of *Cayashinos* remember the communities as *estancias* (ranches), located on the fringes of Hacienda Casca, a huge mining and potato-producing property.[4] *Cayashinos* believe their forebears had not been local to the region but had been brought from the south. Their different origin, local school-teachers now say, is marked by their physiognomy and lighter skin colour as well as by their family names, music and less pure or authentic (i.e. more hispanicised) *quechua* speech. In the teachers' representation, the slur of having been 'Indians of the hacienda' is countered by a claim of superiority, although the idiom in which this superiority is expressed is a deeply racist one – fairer skin and imperfect quechua.

The stories told by *Cayashinos* provide clues about life under the territorialising regime of the hacienda. Apart from defending the northern boundary, residents of the *estancias* had to work as shepherds, transporters, and occasionally mine workers in the hacienda silver mine. They were also dispatched to work in other mines in the region belonging to the owners' family (the Rizo Patron) and business associates. Hacienda rule is remembered as oppressive and impoverishing because it immobilised people. However, stories of the hacienda past are shot through with contradiction, especially with respect to movement. Certainly there was a strong measure of force, for periodically armed guards were sent to check up on the *estancias*. But although administrators deployed and fixed people within the hacienda domain, they could not control surreptitious movement, their slipping away and temporary disappearance. A position on the wild margins allowed Cayashinos to belong to and mediate between two territorialising regimes: the hacienda and the indigenous vertical economy where networks of exchange connected the hamlets to higher and lower resource islands beyond the confines of the hacienda.

Hacienda Casca was expropriated during the Agrarian Reform in the early 1970s and its lands settled on the ex-*feudatarios*, but the Cayash communities failed to win administrative independence. Instead, they were appropriated by a higher ranking settlement, the village of Yanec, which had succeeded in becoming recognised by state and provincial authorities as the *pueblo matrice* (mother town) with administrative authority over subordinate Cayash. To mark its superiority, Yanec kept hold of the precious land titles (and has refused to part with them to the present day). As *pueblo matrice*, Yanec assumed the right to deploy the labour of all those living within its administrative jurisdiction. Thus for Cayash, escape from hacienda rule did not mean the communities could deploy their own labour for their own needs, a situation deeply resented by the community authorities.

It is tempting to interpret Yanec's attempt to use the political openings of the 1970s for its own purposes, to colonise and control the *estancias* of a dismembering hacienda, as the re-emergence of a model of vertical control. But the objective was no longer control over agricultural production, or deployment of labour to work in the resource islands. Yanec demanded labour service to build a road in order to link the village to the main arterial road on the Pampa de Junín that would give access to the high mining centres as well as to Lima and the coast. The authorities of Yanec were also keen to arrange communal work parties (*faenas*) to construct new buildings to betoken its elevated status as an urban place. Here, we begin to see a new imagery coming into play with respect to citizenship, urban-ness and accessibility: the active participation of citizens in the affairs of the state was symbolised by possessing a road. Thus for Cayash, its subordination to Yanec represented a double blow: *Cayashinos* had to contribute labour to a project that would widen the gulf and deepen their inferiority with respect to the *pueblo matrice*. Just as irksome was Yanec's insistence on representing Cayash in all political dealings with outside authorities.

But Yanec's control was fragile. This form of territorialising regime was resented and immediately contested. Following the end of hacienda rule, *Cayashinos* were now more free to move for longer periods to the nearby mining centres, where they could take wage work and find better schooling for their children. As the numbers of educated ex-resident *Cayashinos* grew, they took up the fight for political autonomy and pressed the case for the administrative re-classification of Cayash through the Lima bureaucracy. One can argue that involvement in wage labour did carry an emancipatory potential for it offered improved opportunities for both accumulation and political negotiating. A critical difference then emerged in the political activities of the ex-residents of Yanec and Cayash. Yanec put up little fight. Once Cayash had won higher administrative status it was empowered to take the six hamlets from out of Yanec's control.

Cayash had used the state bureaucracy in Lima as arbiter in its rivalries and struggles but apart from the important exception of primary school teachers, representatives of the state had been absent from the region. Nobody could recall seeing police, military or representatives of the law. The priest came once a year to attend the fiesta of the patron saint and to marry and baptise en masse, otherwise the only outsider was the occasional *ingeniero* who came grudgingly at the request of the communities to survey the track for a possible road. But although the central state was absent, it was constantly invoked. The authorities of Cayash assiduously set out to follow the letter of the law. They were punctilious in filling the offices (*cargos*) in the community institutions and in reporting back to their administrative superiors. They made themselves citizens by responsibly and capably governing and regulating themselves: *agentes municipales* reported to the district council responsible in San Pedro de Cajas; *tenientes gobernadores* travelled to the Subprefect's office in Tarma town, and *presidentes comunales* who superintended the distribution of community land kept contact with the Tarma office of the *juzgado de tierras*. Community authorities travelled frequently to make representations and petitions for improvements, while at home they were left undisturbed to conduct their own affairs. For a time the arrangement seemed to work well, at least in the eyes of the prosperous, politically dominant, families.

The municipal authority was being put under increasing pressure by the early 1980s. There was a mounting clamour, especially on the part of the young, for public works and improvements. *Faenas* were organised (once a month) in each of the communities to repair bridges and tracks, build new community centres and health posts, refurbish schools and churches, pave central squares, and in two of the hamlets install piped water. The central settlement of Cayash succeeded in climbing one rank higher up the administrative hierarchy, being designated a *centro poblado menor* in 1985, but judiciously made little attempt to commandeer the labour of the other hamlets. *Agentes municipales* had to spend even more time travelling to press the Ministry sub-offices to send additional schoolteachers (or complain about those they had) and primary health care personnel as well as press the provincial capital for donations of cash and building materials. But the most pressing demand, to build the 52 km stretch of road between Cayash and the road to Tarma was beyond their capability. The Cayash communities were insignificant in the eyes of provincial and national authorities; the state began to be blamed more vociferously for its neglect; and a different imagination of state-citizen relations was beginning to take hold.

### Re-tracing journeys: the spatial fields of parents and children

Within the communities of Cayash conditions of mobility were not equally shared; some households and individuals were freer to move than others, and some were deployed while others authorized deployment. Differentiation in terms of movement underpinned fundamental aspects of social-economic division. The most prosperous *comuneros* had plentiful pack animals (llamas and mules) and could dovetail the seasonal tasks of the agrarian calendar with regular movement to barter and trade. The men worked several circuits in the course of the year that involved them in different kinds of exchange relations. The journeys are remembered as a set of cardinal points that led east, west, north and south. The journey east to settlements at lower altitudes one day's walk away mapped the principal barter route, where potatoes were exchanged for maize. *Cayashinos* tended to deal with the same families year in year out. In addition to providing maize, calabashes, vegetables and fruit grown in the zone, goods brought up from the lower tropical zone (coffee, yucca, sugar, cane alcohol) were also available for exchange. Only recently had the range of goods expanded at these intermediary points. At the time of the grandparents, the men had been forced to travel further afield to trade and work in the plantations of the tropical lowlands and were remembered as *muy hombre, muy valiente* (very manly, very brave).[5]

The two-day journey west was made several times a year, and the route led up to commercial centres located in the high altitude Pampa de Junín on the main road to Lima. There, potatoes and sheep could be sold to truck traders who supplied the Lima market and offered for sale industrially-made goods: clothes and shoes, plastic goods, construction materials and beer. The journey north led to the end of a dirt road in the Ulcumayo valley and thronged with truck traders at harvest time wanting to buy potatoes in return for industrially-made goods. Of increasing frequency were the journeys south where the most important destination was Tarma town. Even though the round trip was over 100 kms, many men made the journey every month, sometimes every fortnight, spending the night in a cave or with kinsmen at Casca. The *comuneros* brought potatoes, animals, wool and leather for sale in the Tarma market and returned with foodstuffs (sugar, salt, rice), coca, matches, work tools, building materials and animals. Provincial authorities and ministry sub-offices were visited and petitioned. Every man needed to visit Tarma once in a while to acquire his documents of identification, the *libreta militar* and *libreta electoral*, as proof of citizenship.

Travelling took up considerable time for men with pack animals; we estimated that on average one week in five or six was spent away from home. In hindsight, the journeys are recalled not just as necessary or as drudgery but also as an enjoyable and challenging aspect of rural life. When not much was happening, men would decide take a little trip, (*dar una vueltita*), to go visiting. Meanwhile most women stayed at home, responsible for the daily needs of their young children and the animals. Fiestas were occasions for families to visit kin in the other communities; these celebrations lasted three to five days, were accompanied by music from four or five orchestras, and football tournaments (which from the 1980s replaced the traditional bull fights) that might gather some twenty local teams, as well as those sent by the *hijos de Cayash* (sons of Cayash) living in Lima.

From the travel stories several points emerge as to conceptions of space and mobility. First, there had been nothing timeless or traditional about barter and trading circuits. Through barter networks, skeins of social relations were kept intact and these allowed an important measure of security in times of scarcity and want. The survival of barter, as Mayer (2002:158) argues, needs to be understood as a form of protectionism, 'an economic sphere that was separate from the cash sphere, constructed and maintained by the peasantry who, for their own purposes and to their own advantages, tried to isolate it from the cash nexus'. Barter relations had not only endured, they could also be re-invigorated, as was happening in the settlements fringing the tropical lowlands. Although a road connection would clearly be advantageous to connect the communities to Tarma, a single routeway could not compensate for the multiple forms of exchange that had helped assure security and autonomy.

Secondly, *Cayashinos* through their movement produced a spatial field whose co-ordinates were constantly re-inscribed through iterative spatial practice. The spatial field was composed by an east-west and a north-south axis that marked trans-sections across differing ecological zones. They also marked a string of points where kin and families *de confianza* offered shelter, food and goods for exchange. Rights of passage between the points needed to be constantly secured; keeping routeways open demanded vigilance, a constantly moving presence. Not only were spatial connections inscribed through walking, temporal connections were made through the stories and myths that were told about the past; this was a living landscape through which people walked. Without this iterative presence in space and time, as earlier (and later) events proved, spatial fields and rights of passage easily became blocked due to the colonizing, territorialising ambitions of aggressive neighbours. In an Andean context, the making of locality was not only a matter of marking and defending physical boundaries, it demanded the securing of routes, circuits and rights of passage in order that communities remain in touch with a wide resource network. The fact of being off-road might even carry some advantage as it allowed rural people to participate in the external world in their own time and on their own terms.

Third, the travel stories differentiated conditions of mobility at community level. Given gender divisions of labour, young women tended to be more place-bound than the men. However, this did not mean that women felt immobilised; as women got older many left to take charge of selling Cayash's *papas de color* in the markets of Tarma, Junin and La Oroya. Mobility also distinguished socio-economic status and a threefold division could be discerned amongst the households. In the first place, a polarisation had emerged between prosperous households with pack animals (which produced surpluses, hired labour and were more mobile) and impoverished households (without pack animals whose menfolk were more tied to place). Some 10% of the hamlets' population were estimated to belong to this category of rural poor. They did not possess sufficient resources to work community land and hired themselves out as labourers (*peones*); their diets were more meagre, their clothing threadbare and some still went without shoes. There was also a third group: the *Cayashino* families who straddled wage work in the mining towns with a presence in the communities to plant and harvest. Outside these groups but still linked in some ways to the communities were the ex-residents in Tarma, La Oroya and Lima and on the outer fringes, individual *Cayashinos* who had settled in the US, Panama and Argentina and who were blamed for never making the journey back to Cayash even at fiesta time.

Differential status dovetailed with distinctions drawn along lines of 'race'. Families whose mobility was more constricted were portrayed as both poor and more Indian, a view that had become increasingly pronounced amongst the youngsters who had attended school and who picked up the racist discourse of the schoolteachers (Wilson, 2000). The conflation of mobility and 'race' also underlay the way in which people from one hamlet/community spoke of the others. By labelling people as *mas humilde, mas cerrado, menos despertado* (more humble, more closed, less awakened), as opposed to *mas tranquilo, mas democratico, mas abierto* (more peaceful, more democratic, more open), an association was made between movement, state of awareness, and receptivity to change that brought to mind the old distinction between Indian and *mestizo*.

Finally, we come to division in terms of generation. Through the lens of movement we can discern the widening gulf between life worlds of parents and children. In exacerbating this division, the parents had been unwittingly to blame. Expecting respect and complete obedience, parents had felt no qualms at sending their children away for education, this being seen as a new form of resource niche that could be colonised. After all, many from the parents' generation had been sent off to work at a very young age to earn money, for example on the potato-producing haciendas of Huasahuasi, to pay their parents' financial obligations at fiesta time. Deployment in space had become a family matter where the adults dictated and governed the mobility of the young. But with the expansion of secondary education, a triumph of the reformist military government of the 1970s, sending young children to the towns for secondary education came to have unforeseen consequences.

During the 1980s, increasing numbers of children from peasant communities were sent to secondary schools in town. Their experiences can be illustrated by the life story of one youngster from Cayash whose father had been a community leader. Daniel (born in Cayash in 1970) was the oldest of nine siblings. He began helping out with the family's animals as a four year old, and then he joined a class of 45 pupils at the local primary school. When he reached 11 years, his parents sent him along with his older sister to Tarma so they could both attend secondary school. Since no close relative then lived in Tarma, the children were lodged in a rented room and had to fend for themselves. Their father visited them every month and brought food. Daniel remembered the move to Tarma as an intensely painful experience; he recalled the homesickness suffered by children uprooted from the countryside, their sudden loss of family, and their growing awareness of the disdain with which they were treated by pupils and teachers from the town. Some broke down and ignominiously were returned home.

Youngsters from peasant communities had to confront the insidious racism of everyday life in the predominantly mestizo towns. Although it was children from the most prosperous, the whitest and least Indian families, who were given the opportunity of secondary education, they felt themselves treated differently, as uncultured Indians. In the early 1980s, the youngsters found themselves in a highly charged political atmosphere. In Tarma, as elsewhere, this followed in the wake of the heavy military repression of the radical teachers' union, which had pushed political activity underground and given an impetus to Maoism and to a politics of violence. Secondary school children were targeted by political ideologues, especially from the clandestine Maoist party, *Sendero Luminoso*. Many youngsters were open to a political ideology that offered them 'the truth', the hope that through their militancy they would earn status, respect and authority, and the promise that as disciples of a learned leader in a tightly organised, highly authoritarian structure they would find social moorings that their non-comprehending fathers could not provide. The children had been deployed into a hazardous outside world, one in which the quasi-religious ideology of political violence could all too easily colonise them.

In the early days, Daniel returned to Cayash during the school holidays to help his father but latterly he stayed away to earn money for his keep, and release his father from this burden. Absent from home from a young age, Daniel never took on posts in community organisations. Instead of working the land or trading, he tried his hand at a range of urban jobs moving between Tarma, Lima and Pichanaki (a town in the tropical lowlands) where he learnt the rudiments of baking, machine repair, car mechanics, accountancy and Chinese cooking. Throughout the years Daniel continued to identify himself as a *Cayashino* and his parents continued to remind him that social advancement should not be bought at the cost of neglecting his wider responsibilities to defend his community. But like many youngsters he was caught in a dilemma. Their meagre education did not lead to jobs in the professions other than school teaching. And at home, despite the apparent prestige that an education gave, their advice was often adamantly rejected by their fathers. Daniel felt an obligation to return home and in 1989 he found temporary appointment as a teacher in the new secondary school that opened in Cayash that year. A few months later, *Sendero* invaded.

## The violent imposition of an overarching regime

The period of autonomy in Cayash came to a violent end in 1990. When the leaders of *Sendero* decided to shift the main arena of conflict to the Central Andes, Cayash was named as a support base to serve the needs of some 200 militants. Under *Sendero* control, municipal and community authorities were disbanded, the whole population was prohibited from moving without authorization, and all documents of identification were confiscated. The few families who slipped away forfeited their lands, livestock and household goods that were distributed amongst the poor. *Sendero* militants took over the schools and organised frequent political meetings in the churches and central squares. The message they preached was that the aristocratic state had deliberately neglected to attend to community needs; the state had treated them with disdain and transformed them into *las comunidades mas olvidadas* (the most forgotten communities). Since the state was cheating and abusing the people, it was their duty to take up arms and wage a people's war. Daniel's father recalled his confusion in the face of these allegations; he considered himself a 'simple peasant', and refused to pronounce on whose version of the truth was the right one. But he did realise that the state would send an army and that there was good reason to be afraid.

After a year of *Sendero* occupation, the military invaded and this provoked the displacement of the entire population. Occasional reports then appeared in the regional and national press portraying the Cayash hamlets as hotbeds of terrorism. The military hunted down, gaoled and tortured many of the young, educated *Cayashinos* (including Daniel). Branded as terrorists and accomplices, *Cayashinos* faced a long hard struggle to clear their name. Displacement turned out to be another form of containment for most people were fearful of travelling and of being apprehended at the military road blocks. While the men went to ground, women travelled to try find family members who had disappeared. Eventually, municipal and community authorities were re-constituted in Tarma town and a pact of reconciliation was reached 1993 that put some break on military harassment. Resettlement slowly got underway in 1994 but without the assistance of the state, or donor agencies (Wilson, 1997).

The dream of recuperation became closely connected with two demands: re-claiming citizenship (in order to lose the label of subversive) and building a road (in order to lose the label of ignorant – and violent – Indian). In hindsight, one can suggest that by focussing political activity on building a road, municipal authorities tried to foster a shared, collective view, one that would plaster over the wounds of political division. One is struck by the intensity of political activity, the energy expended in making constant representation and appeal to higher administrative authorities. Petitioning and prevailing upon central and local authorities were integral to the peasantry's concept of citizenship. They argued that the state had a moral obligation to provide a road to allow people to home return with dignity; it was a matter of social justice. But, this was a line of argument that no longer found favour in the neo-liberal state. The rationale for state support for road building was not nebulous reasons of social justice or even poverty alleviation; increasingly the state was defining road-building as a strategy to give capitalists access to the agricultural frontier and increase production.

Under the Fujimori government, a road plan for Cayash was drawn up and approved by the Ministry of Transport in Lima and a three year construction programme begun. The project document made much of the development potentials the road would open up. A road would allow: (i) the substitution of traditional systems by new modern techniques of cultivation and livestock raising; (ii) provide access to credit; (iii) allow greater control by *organismos* of the state, (iv) offer capitalists access to known reserves of mineral wealth; (v) facilitate an intensive training of peasant producers; (vi) allow an increase in productive infrastructure; and (vii) make possible the establishing of a new socio-economic structure in the region. The road project document suggests a picture of colonisation from the outside, far removed from what the struggling community authorities of Cayash had in mind. The future envisaged in the document is one where peasant knowledge and experience is discredited and debased; left unstated is the aspiration of the large potato producers of Huasahuasi to push out their frontiers of production and appropriate the lands of Cayash. Thus the road may well put at risk the rights and entitlements of Cayashinos to their lands and resources.

Cayash and its road are being drawn onto the national map; a new territorialising regime appears to be coalescing. However, one is justified in fearing that once road-building is completed, if it is not accompanied by a resurgence of community action then the road will herald an invasion of land-claiming outsiders from neighbouring districts and an undermining of livelihoods and way of life. But the odds are stacked against communities like Cayash. The struggle to renew the continuity of rural life worlds demands far more than the partial re-settlement of a handful of hamlets. It seems highly unlikely that the complex barter networks and multiple strands of social relations held in place through the movement of peasants with their pack animals will ever be repaired.

Recent history in the Peruvian Andes has been characterised by a kaleidoscoping, a more rapid succession of territorialising regimes, each producing its own conditions of mobility. There have been worlds of difference between them. One has been the iterative marking out of community territory through its route and connections to the outside world, walked and talked in accordance with the seasons. A different variant was seasonal movement between communities and the mines. Different in scope, scale and content was the mobility of the educated young. Seasonal or periodic though this movement might be (as temporary jobs were taken in various places in different seasons), the life worlds became attenuated and broken off from the community of their birth and belonging. In this situation, substitute communities were found such as amongst the Maoist cadres or in the ranks of the *evangelistas* (of the protestant churches) – or in both. Military invasions had brought stasis, when people were either penned up in their communities or immobilised after violent displacement.

My point is that the story of Cayash suggests how differentiated mobilities both interweave and separate social actors over time. While one can point to the perpetuation of some discourses that are shared, those underpinned by the Andean model of vertical control, celebrated and perpetuated through a particular form of mobility, are being made redundant. They are shared neither by the younger generation nor by those opting to remain displaced in the towns. In contrast, defence of community now draws on a repertoire of social meanings that has less to do with actual spatial practice, and a great deal more to do with geographies of imagination. The ideal of community is not dead, as seen by the vibrant interest in the road, and astonishing willingness on the part of far-flung members to contribute financially to the restoration of Cayash. But for the large majority who stay away, the community of Cayash is becoming 'dis-placed'; it has dissolved into heterotopia, and defined as a place of tradition and nostalgia, held safe in the mind's eye from the discordant experiences of invasion, reprisal and stigmatisation or feared experiences of painful return.

# Notes

1 For recent appreciations of Murra's work, see Harris, 2000 and Mayer, 2002.
2 See for example Murra, 1975, and the essays in Alberti and Mayer, 1974, and Lehmann, 1982.
3 Bradby, 1982:111–112. Recently, Enrique Mayer (2002) has taken up the debate again, stressing how Andean socio-political organisation rested on high levels of inter-household collaboration, institutions through which communities could control their members, and relations of cooperation ritualised and subject to rules, sanctions and social pressures.
4 According to the population census of 1876, there were 91 people resident on hacienda Casca; by 1940, according to the census, there were 153 people, or 31 families.
5 Smith, 1991:89, notes that in Huasicancha owing to labour shortage, instead of each household being involved with barter relations in the tropical lowlands, they had assigned a group of young men to the task.

# References

Agnew, J. and S. Corbridge (1995) *Mastering Space: Hegemony, Territory and International Political Economy*. London and New York: Routledge.

Alberti, Giorgio and Enrique Mayer (eds) (1974) *Reciprocidad e intercambio en los Andes peruanos*, Peru Problema 12, Lima, Instituto de Estudios Peruanos.

Allen, J. (1999) 'Spatial assemblages of power: from domination to empowerment', in D. Massey, J. Allen and P. Sarre (eds) *Human Geography Today*, pp. 194–218. Cambridge: Polity Press.

Bourdieu, P. (1991) *Language and Symbolic Power*. Cambridge: Polity Press.

Bradby, B. (1982) ' "Resistance to capitalism" in the Peruvian Andes', in D. Lehmann (ed.) *Ecology and Exchange in the Andes*, pp. 97–122. Cambridge: Cambridge University Press.

Gupta, A. (1995) 'Blurred boundaries: the discourse of corruption, the culture of politics and the imagined state', *American Ethnologist* 22(2): 375–402.

Hansen, T. and F. Stepputat (2001) Introduction, *States of Imagination: Ethnographic Explorations of the Postcolonial State*. Durham: Duke University Press.

Harris, O. (1995) 'Ethnic identity and market relations: Indians and mestizos in the Andes', in O. Harris and B. Larsen (eds) *Ethnicity, Markets and Anthropology: At the Crossroads of History and Anthropology*, pp. 351–90. Durham: Duke University Press.

Harris, O. (2000) *To Make the Earth Bear Fruit: Ethnographic Essays on Fertility, Work and Gender in Highland Bolivia*. London: Institute of Latin American Studies.

Herzfeld, M. (1997) *Cultural Intimacy: Social Poetics in the Nation-state*. New York/London: Routledge.

Kay, C. (2001) 'Reflections on rural violence in Latin America', *Third World Quarterly* 22(5): 741–75.

Larson, B. 1995, 'Andean communities, political cultures, and markets: the changing contours of a field', in (eds) Larson and Harris, *Ethnicity, Markets and Migration in the Andes: At the Crossroads of History and Anthropology*, Durham, Duke University Press.

Lefebvre, H. (1991) *The production of space*. Oxford: Blackwell.

Lehmann, D. 1982, 'Introduction: Andean societies and the theory of peasant economy', in Lehmann (ed), *Ecology and Exchange in the Andes*, Cambridge, Cambridge University Press.

Mayer, E. (2002) *The Articulated Peasant: Household Economies in the Andes*. Boulder: Westview Press.

Murra, J. 1975, *Formaciones economicas y politicas del mundo andino*, Lima, Instituto de Esudios Peruanos.

Radcliffe, S. (2001) 'Imagining the state as a space: territoriality and the formation of the state in Ecuador', in T. Hansen and F. Stepputat (eds) *States of Imagination: Ethnographic Explorations of the Postcolonial State*, pp. 123–45. Durham: Duke University Press.

Roseberry, W. (1994) 'Hegemony and the language of contention' in G. Joseph and D. Nugent (eds) *Everyday Forms of State Formation: Revolution and the Negotiation of Rule in Modern Mexico*. Durham: Duke University Press.

Slater, D. (1998) 'Re-thinking the spatialities of social movements: questions of (b)orders, culture, and politics in global times' in A. Alvarez, E. Dagnino and A. Escobar (eds) *Cultures of Politics, Politics of Cultures: Re-visioning Latin American Social Movements*, pp. 380–401, Boulder: Westview.

Smith, G. (1991) *Livelihood and Resistance: Peasants and the Politics of Land in Peru*. Berkeley: University of California Press.

Smith, N. (1992) 'Geography, difference and the politics of scale', in Doherty, Graham and Malek (eds) *Postmodernism and the Social Sciences*, London: Macmillan.

Wilson, F. (1997) 'Recuperation in an Andean region', *European Journal of Development Research* 9(1): 231–45.

Wilson, F. (2000) 'Representing the state? School and teacher in post-Sendero Peru', *Bulletin of Latin American Research*, 19: 239–253.

Chapter 9

# The Making of Globalized Everyday Geographies

Benno Werlen

## Introduction

For a long time the clarification of the society-space nexus was discussed only as a problem of theoretical conceptualization within human geography, mainly in social and political geography. Today it has become both a crucial sociological and a political problem. It is obvious that under these conditions, human geography assumes a new specific political relevance for contemporary societies as well as a heightened theoretical significance for the social sciences. To make use of its potential we need further adjustments of ontological and geographical research.

Today's transformations of the geographical and socio-cultural conditions of everyday life are occurring at a number of levels, from the changes in political maps to technological inventions in production, transport and communication. The last of these is largely responsible for the globalization of nearly all domains of our day-to-day lives. Still, both aspects – the constant changing of political maps and the process of globalization – involve new combinations and recombinations of the 'social', 'cultural', 'economical' and 'spatial'.

My argument is that, for a better understanding of these forms and all other processes of 'regionalization' and the constitution of any form of socio-spatial relations, we must not base our analysis on the spatial aspects of social conditions *per se*, but on those activities that constitute the socio-spatial relations. The starting-point should not be space or spatiality, but what I have elsewhere called 'everyday regionalizations' (Werlen 1997) and forms of 'world-binding' (Werlen 1999).

The best way to grasp processes of regionalization is the analysis of practices, not primarily of spatial transformations. This categorical shift involves a shift from 'space' to 'action', or from what I call a 'geography of objects' to the 'geographies of subjects'. With this shift, what is referred to as 'globalization' is a number of processes that constitute spatial relations through interactions of various kinds and at various levels. All forms of regionalization are nothing but the result of new forms of everyday geography-making in the course of everyday activities.

Accepting the perspective that I am going to introduce in this chapter also involves a new understanding of 'regionalization'. It is no longer seen as the constitution of container spaces and societies, or spatial demarcations and classifications. It has to be understood rather as a process of what can be called '*world-binding*' by agents. In this process of 'world-binding', the combination of the social and the spatial takes very different forms. At the same time it is one of the most prominent characteristics of globalization.

This conceptualization of regionalization also leads to another understanding of 'region' as proposed by Gregory (1982), Thrift (1993) and Paasi (1991) in the context of the 'new regional geography', but also as in the more recent debate by Allen, Massey and Cochrane (1998) or Scott (2001). 'Region' will no longer be understood as socio-spatial 'fact', but rather as an outcome of world-binding and as an element of life-worlds. Spatial connotations are one aspect of these processes, but not the core element.

One of the important conditions and implications of this perspective is that spatial aspects – depending on the context of action – are both gaining and losing their relevance: 'A bizarre adventure happened to space on the road to globalization: it lost its importance while gaining in significance' (Bauman 2000:110). Secondly, globalization and regionalization can no longer be seen as conflicting tendencies. Regionalization has to be understood as an expression of world-binding, as an implicit aspect of globalization. The starting-point for this approach is an action-centred understanding of space.

### From 'space' to 'action'

For a better understanding of the significance of space for social processes, we do not – as is very often claimed – need a spatial turn; on the contrary, we need a better understanding of the spatial implications of social actions (Werlen 1983, 1987, 1995, 1997, 2000). We have to clarify why 'space' is losing its importance by 'gaining in significance'. If we want to have access to the globalizing logic of everyday geography-making, then we can no longer think about social realities in totalizing spatial categories. If the spatial representation of the social world was not too problematic in the context of traditional life-forms and in the age of the nation-state, it becomes more problematic in the age of globalization. Space has to be thought of as an aspect of action, not as a container of actors and actions.

If we begin to look at the geographical world from an 'action-centred perspective', and discard space as a starting-point in itself, the embodied subject, the corporeality of the actor in the context of specific subjective, socio-cultural and material conditions come to the fore. We then adopt a perspective that emphasizes subjective agency as the only source of action, and hence of change, while at the same time stressing that the social and material world also shapes social actions, while the latter produce and reproduce such conditions.

In fact, any concept of 'space' can only provide a pattern of reference with which problematic and/or relevant material entities that bear on action can be reconstituted and localized. Given that the subject is embodied, these material patterns are of course significant in many actions. But since they are not the only significant factor in action, actions can be neither analysed nor explained by them. 'Space' and materiality become meaningful in the performance of actions only. Of course, as far as spatial patterns are the outcome of actions, the meaning of actions is incorporated in them. But again, only the incorporated meaning has social significance for action.

By consequence, 'space' can determine neither actions nor the agents' frame of reference. Rather, space, as a form of reference, is itself the product of social actions, particularly communication.

If one accepts this postulate, one must first clarify the ontological status of 'space'; and secondly, one needs to specify the relevant kinds of meaningful performances and types of social conditions. I will now discuss this from an analytical perspective, and will then turn to the everyday level.

If 'space' were an object, we should be able to indicate the place of space in the physical world. This is obviously impossible. Space does not exist as a material object, or as a consistent theoretical/political object. It is instead – and this is my thesis – a formal and classificatory concept, a frame of reference for the physical components of actions, and a grammalogue for problems and possibilities related to the performance of action in the physical world. 'Space' is a formal frame of reference because it does not refer to any specific topical aspect of any object. It is 'classificatory' because it enables us to describe a certain order of objects with respect to their specific dimensions.

With this conceptualization of 'space', we can now make clear why 'space' takes on so many different significations in everyday actions. I propose to distinguish three main types of interpretations in formal and in classificatory respects: a rational, a normative and a communicative interpretation. The main reasons for the different interpretations, and correspondingly the different constitutions of 'space', lies in the fact of specific interconnections between the body and other material aspects of situations. The different constitutions of 'space' are expressions of different interpretations of its formal and classificatory dimensions.

| Action | formal | classification | examples |
|---|---|---|---|
| **rational** | geo-metric, absolute | calculation | land market; location theory |
| **normative** | geo-metric, body-centred | normative prescription | nation-state; back/front region |
| **communicative** | body-centred | relational signification | regional/national identity; regional symbol |

**Figure 9.1   Actions and constitutions of spaces**

*Rational actions, space, and the economic dimension*

For rational actions, the social constitution of space is based on the metric aspect. This enables calculative orientation and ordering. Rationality and the geo-metric are tied together closely and both are core expressions of what Max Weber (1980:635) called the de-mystification of the world. This is associated with the formal understanding of space and is the main pre-condition of the rational calculation of spatial extensions.

As a result, a precise cartographic representation becomes possible, along with the emergence of a capitalistic land market. Together with the invention of the mechanical clock, the formalized space concept is the prominent basis for industrial capitalism. In fact, the rationality of the domination of nature through modern technology is to a certain extent based on the rational space concept, as well as the idea of rational land-use planning by state institutions.

*Norm, space, and the political dimension*

The interrelation of action and space in a norm-oriented context leads to *territorializations*. The formal aspect involves a geo-metric appropriation of extensions in a body-centred way. The classificational aspect refers to the relations among body, material context and normative prescription in the form of: 'Here you can do this, but there you must not' etc. This kind of territorialization as the prescriptive form of regionalization governs the inclusion and exclusion of actors and resources.

At the personal level this includes the differentiation between back and front regions (Goffman 1969; Giddens 1984). But the most prominent combination of norm, body and space is certainly the nation-state and its territorial binding of law, courts, territorial organization of bureaucracy, surveillance and control of the resources of violence by police and army.

*Communication, space and the cultural dimension*

In communicative actions, spatial orientations are also predominantly body-centred. The body is the functional link between experience (stock of knowledge) and meaning, and is the operational basis of subjective action. The attribution of meaning to things depends on our knowledge of them and their roles in our actions. The symbolic constitution of meaning is therefore the result of the interplay of knowledge, attribution and materiality. This is one of the dimensions of symbolic everyday geography-making: the symbolic appropriation of material things. These processes are also typical of the constitution of national identities in the course of the building of nation-states, and of the discourses of nationalist and regionalist movements.

Furthermore, the body is crucial to the constituting processes of intersubjectivity. If a subject is learning intersubjectively valid rules of interpretation that exist within certain socio-cultural surroundings, it is necessary for him/her to verify his or her interpretations and evaluations. This means that the constitution and application of an intersubjective meaning-context depends on the possibilities of testing the validity of allocations of meaning. Certainty about intersubjectively valid constitutions of meaning is primarily possible in the immediate face-to-face situation. It is one of the distinctive features of the globalizing mechanisms that most of the communications are mediated, lifted out of the immediate face-to-face situation.

## Globalization and subjective action

In fact, the field of information and communication is the most prominent one in the globalization process, enabling new forms of economic transactions, organizations of production and consumption. The disembedding mechanisms are grounded in the history of modernization. These mechanisms are at the core of the transformation of the space-society-nexus and make it obvious that space is neither a cause nor a container, nor in fact material at all.

The traditional life-form, which can be characterized by the fact that most frames of reference for action were spatially embedded, could lead to such an understanding of space. But the spatial clustering and embeddedness of traditional social life-forms has been replaced more and more by global interconnections and disembedding mechanisms. With the growing power of disembedding mechanisms it also becomes obvious that all spatial aspects are nothing but frames of reference constituted in respect of the type of action that needs to be (or is) performed.

The growing power of disembedding mechanisms is illustrated by the globally observable cultures (Featherstone 1990; Albrow 1996; Beck 1998), lifestyles and life-forms (Schutz 1982; Shields 1992; Chaney 1996) that are very often linked to a specific generation. The actual and potential reach of actors has been extended into a global dimension (McLuhan 1995). The most important disembedding mechanisms in this respect are money, writing and technical artefacts (Curry 1996; Strohmayer 1998). Means of transportation enable a high level of mobility.

Together with individual freedom of movement, this produces a mix of formerly locally rooted cultures. This multicultural medley, combined with global communication systems, enables a diffusion of information and information storage that are not dependent on the corporeal presence of the actors. Of course, face-to-face interaction still exists as an important situation of communication, but the most substantial part of communication is now mediated.

'Space' and 'time' are emptied of fixed signification, or at least separated from them (Durkheim 1998; Sack 1980). The signification of things is much more the result of a recombination by the subject, depending on the performed action. What a thing signifies is no longer taken as a quality of the thing itself, but is rather attributed to it, and the content of the attribution depends in principle on what the subject is doing or wants to do. Therefore, the 'When' and 'Where' of social activities are things to agree upon, subjects of agreement, and do not depend on fixed, pre-given meaning-contents of social activities. The place of traditionally fixed meanings is taken over or replaced by rationally and institutionally determined regulations, open to communicative revision.

Of course the transformative power of the disembedding tools only becomes effective in the life-world if the subjects integrate them in their courses of action. Because they do so, the transformations of their own day-to-day life are so radical that the use in the social sciences of the 'life-world' concept itself has to be reconsidered. Luhmann (1986:179) argues that under such conditions the life-world concept – as it is used for example in the work of Schutz/Luckmann (1974) and Habermas (1981) for the understanding of contemporary cultures and societies – became nothing but a confusing metaphor.

I will argue that this confusion lies in the fact that the basic distinctions in Husserl's definition of 'life-world' have not been transferred with enough accuracy to the social sciences. To discuss this, I will briefly reconstruct the reception of Husserl's key concept in the social and cultural sciences. For Husserl (1989:355) 'the life-world is the natural world – in the attitude of natural life we are living functioning subjects together in an open circle of other functioning subjects'. Schutz/Luckmann (1974:3) identify this with the realm of reality, that the 'everyday life world is to be understood as that province of reality which the wide-awake and normal adult simply takes for granted in the attitude of common sense. By this "taken-for-grantedness" we designate anything we experience as unquestionable; every state of affairs is for us unproblematic until further notice'.

Now, if we interpret this definition to mean that the familiar, non-questioned sections of the pre-given world are those that are spatially close, then it loses its analytical potential for the social-scientific investigation of social-cultural worlds in the situation of globalization. And this is certainly the case for the interpretation that Schutz/Luckmann applied by linking 'certainty' with the various scopes of action, from those 'within immediate reach', which offer 'the fundamental test of all realities' (Schutz and Luckmann, 1974:42), to the most distant 'zones' with a high degree of uncertainty. Consequently, 'man within his natural attitude is primarily interested in that sector of the world of his everyday life which is within his scope and which is centred in space and time around himself' (Schutz, 1962:222). Habermas too (1981:226) links 'life-world' – through the mechanisms of social integration – to the insiders' perspective of co-present group members.

But Husserl's definition implies no immediate space nexus. One life-world is characterized by a certain 'attitude' and a certain topical 'horizon'. The *first* of these – Claesges (1972:86) calls it a 'grounding function' (*Boden-Funktion*) of the life-world concept – makes it possible to clarify the difference between a natural (day-to-day level) and a theoretical attitude (scientific level). The *second* refers to different topical horizons of interest (*Horizont*), in which subjects live in their courses of action in their day-to-day lives. The interests of action limit specific 'Sonderwelten' (Husserl, 1976:194). Luhmann is only right to talk about confusion if one fails to distinguish clearly between the two concepts. This is especially the case if the two are not seen as different accesses to the socio-cultural world, but as the result of an ontological distinction.

It is very important to see that according to Husserl the '*Boden-Funktion*' refers to an epistemological dimension, and the '*Horizont-Funktion*' to the empirical aspect of day-to-day activities. The importance of this distinction becomes obvious first of all in a globalized world. Under these circumstances the familiar is no longer strictly tied to the local community as was/is the case in traditional life-forms. Topical life-worlds' reach are not free of doubt and uncertainties. Late-modern subjects can live in 'Sonderwelten' with global reach. Life-world studies in a geographical perspective have to take this into account.

This definition of life-world has very strong implications, particularly for an action-centred understanding of the interrelation of globalization and regionalization.

**New understanding of regionalization**

Because of the disembeddedness of late modern life forms, spatially homogenous societies and cultures hardly exist any more, and the contours of spatial differentiation are becoming more and more indistinct. Consequently, spatial characterizations of the socio-cultural are losing their precision and validity. The so called 'third world', for example, can also be found in New York or Paris, as the 'first world' can be found in Nairobi, Kinshasa, La Paz or Bangkok. What seems to be a rather simple and well known phenomenon is just the surface of a deep transformation of social and cultural realities, which in some respects are becoming elements of 'placeless spaces' (Castells 1996).

An adequate contemporary representation and understanding of the interrelation of society and space has to be able to take the new ontology of the social world into account. What is needed is a scientific approach that is not alienated from the actual life-forms produced by the actors, despite the ideological discourses of fundamentalist, nationalistic and regionalist movements. To grasp the day-to-day processes of globalization is, after all, a very crucial project of social science, beyond any false reification.

One other implication of the disembedding mechanisms for people's actions certainly is, and will continue to be, that the most important social differences will increasingly lose their spatial form. Already today we can observe the very dramatic differences in incomes within the smallest spatial contexts. There is a lot of evidence that these tendencies are unlikely to lose their transformative power (Held 1999), and that they will only be more accentuated in the future. If David Harvey (1996:429, 2000) postulates that 'uneven geographical development' should still be the main concern of geographical research, the question that has to be raised is how well the growth of differences inside these geographical units can be recognized.

Traditionally, 'regionalization' in geography is defined as a scientific praxis of spatial classification. At the everyday level, 'regionalization' often signifies a process of political appropriation and/or delimitation. In either case, the constitutive idea is based on the spatial delimitation of the 'region'. Borders of everyday political regions normally consist of symbolic or material markers. But 'physical markers' are, in their social aspect, nothing but material representations of the symbolic delimitations of normative standards. Material conditions are therefore not social constraints – only the social norms are. Spatial aspects are consequently neither causes of nor reasons for actions in and of themselves. They exist socially only in the way and to the extent that they are mobilized as means of categorization and symbolic representation in actions.

Consequently, the central role of regionalization is, *first*, not spatial delimitation. It is rather the process of appropriation. Putting this in a more general context, we can understand regionalization as a form and a process of what I have called 'world-binding', which means a praxis of allocative appropriations of material objects, authoritative appropriation of subjects as well as symbolic appropriation of objects and subjects.

'World-binding' as the core element of any kind of regionalization process can thus be defined as a praxis of re-embedding: *to bring the 'world' within the reach of actors through their culturally, socially and economically uneven capacity to hold sway over spatial and temporal references*. Social control of spatial references enables the direction of the subjects' own actions or of other actors' (possessive) practices. This implies the threefold (allocative, authoritative and symbolic) appropriation of goods, persons and objects/places over distances.

*Secondly*, it is therefore not the production of 'spaces' that is of central interest, but the use of spatial and temporal dimensions for the different types of appropriation. *Thirdly*, globalization can be understood as a process of appropriation, as a set of specific forms of '*world-binding*'. The globalization of the life-worlds is consequently the expression of the multiple combinations of world-binding on the basis of a wide range of individualistic choices. These decisions are of course limited by the capacity to control material goods (nature, artefacts etc.), the allocative resources, the capacity to control people (authoritative resources) and the capacity to control symbolic attribution and the management of meaning.

Of course, even the processes of building nation-states and the subsequent nationalization of life-forms were and are forms of world-binding. The nationalization of life-forms can be understood as the rationalization of traditional life-forms along the core dimensions of modernization. The disembedding of traditional life-forms led to a re-embedding through rational territorialization. The most significant expressions of this process were the transformation of the relations of production and exchange (capitalism), the transformation of technologies of production and communication (industrialism) and the emergence of the powerful apparatus of bureaucracy (bureaucratization) for the coordination and control of human actions over long temporal and spatial distances.

A central aspect of the history of modernization is the processes of territorialization and therefore of regionalization at the level of nation-states. Religious-mythological embeddedness has been replaced by bureaucratic-institutional forms of re-embedding. The history of nation-states may be the most prominent expression of this process. In the age of late modernity these basic principles are transformed, especially at the economic and cultural level. The modern principles of territorialization and regionalization are evaporating as a consequence of the growing power of disembedding mechanisms. One of the striking implications of this is the globalization of life-worlds.

## Enacted, globalized and re-embedding geographies

What types of everyday geographies or geographical life-worlds under globalized conditions can hypothetically be identified? With respect to action and structuration theory, three main types of everyday regionalizations of life-worlds (Werlen 1997) can be thematicized (Figure 9.2):

| Main types | Sub-types |
|---|---|
| **productive – consumptive** | Geographies of production<br>Geographies of consumption |
| **normative – political** | Geographies of normative appropriation<br>Geographies of political control |
| **informative – significative** | Geographies of information/knowledge<br>Geographies of symbolic appropriation |

**Figure 9.2   Main and sub-types of globalized everyday geographies**

*Globalized geographies of production and consumption*

In analysing everyday geographies from an action perspective, the first question we need to address is this: how do subjects produce geographies by situating the objects of particular activities, and how do they create and maintain a certain order of objects by means of consumption? This initially orients the analysis towards the less complex forms of the regionalization of the life-world in the productive-consumptive domain, centred on the economic dimensions of everyday reality.

The productive side is most obvious in the form of decisions on the location of productive activities, the subsequent establishment of the spaces of action for working people and the patterns of commodity flows as inputs to the production process (Scott 1998). This corresponds to the analysis of the global organization of the capitalist production regime. It is, of course, not only as producers that we make geographies; we also do so as consumers (Certeau 1988; Shields 1992; Jackson/Thrift 1995; Miller 1995; Crewe/Lowe 1996), even if these forms are more implicit. In globalized life-conditions, our personally defined lifestyles have strong implications for the structuration of the world economy. All in all, the aim of the analysis of this main type of regionalization is not to explain spatial patterns; it is much more to reconstruct the global implications of our locally based life-forms.

Each decision in the context of a certain style of life and consumption may be very trivial. But with the growing commercialization of everyday life the number of decisions of this kind is *in the first place* increasing; and *secondly* their global implications are growing proportionately. Despite their triviality, these decisions become real authority in the formation of a (world) economy. Consequently, consumption is no longer just the determined consequence of production. In globalized conditions consumption is much more an active process; that is, it constitutes – in a silent and invisible way – a mode of production in itself: the art of using products leads to the realization of a style (de Certeau 1988:13), establishing a 'symbolic capital of representation' (Bourdieu 1994:213f).

In this context, 'shopping' is at the same time both an act of representation and is conceived as an event. Hypothetically, one can claim that subjective consumption follows a mentally represented script. These scripts are influencing the kinds of production as well as the *mise en scène* of the products by the producer. Consumption becomes an important part of the construction of self-identity, and 'shopping for subjectivity' (Langman 1992, 40) is one of its main features. All in all, globalization is maintaining and accelerating the culturalization of the economic.

*Normative and political geographies*

A second domain of everyday social geography concerns the normative-political *interpretations* of 'zones of action' or territories. Again, the starting-points are the body-centred regionalizations of the 'front regions' of social presentation (i.e. stage, performance etc.), and the 'back regions' (Goffman 1969) of social concealment (i.e. intimacy, shame etc.) and their differentiation with respect to age, sex, status and role. In addition, one might consider the territorial regulation that involves the inclusion and exclusion of actors by property rights, political/legal definitions of nation states and citizenship rights. Focusing on the interpretation of the subjects is a consequence of the shift in perspective from space to action, the making of everyday geographies.

These types of everyday social geography are linked to the authoritative control (Agnew/Corbridge 1995; Guibernau 1996) of people by territorial means (Philo 1989; Paasi 1991; Sofsky 1997), as in 'geographies of policing' (Fyve 1991), and specific ways of controlling the instruments of violence.

If we start from the premise that all non-natural geographies are produced and reproduced by human practices, then we can gain access to the crucial role of space in the reproduction of authoritative resources and power, i.e. of the political everyday geographies. Power can thus be identified as a dimension of action involving the use of material goods as well as the control of other actors. To have power over and control of space means to have control of subjects by controlling their bodies. The interrelation between power and space can now be identified as an interrelation between power and the body through normative appropriation. All forms of territorialization and political control are based – and this is a hypothesis for subsequent empirical research – on this interrelation.

The interrelation specified above is fundamental to the nationalization of life-worlds in the context of the modern nation-state and the prescriptive appropriation of territories of any kind, leading to (national) container societies. In any case of economic action, the principle of territoriality is applied to the tokens of exchange (money). These relations are in turn part of the constitution of a new 'society-space' nexus, whose specific features are both rational territorialization and the containerization of the political and legal order through bureaucratic organization.

The containerization of the cultural spheres is based on the territorialization of education and information systems. The emergence of national (high) languages is the cultural parallel to the emergence of a single national currency at the economic level.

A very important component of the making of everyday political geographies is the activities of regionalist and nationalistic movements which aim for a new political geography – very often by simply looking for a new container – and the different forms of regional and national identities on which they are based. The increasing importance of regional and national identities can be understood as a consequence of the rise of new disembedding mechanisms.

We can start from the premise that the disembedding mechanisms that extend the field of decision-making possibilities lead to considerable insecurities at the personal level. In these conditions the need for stabilizing identities grows. In this sense 'regionalism' can be seen as compensation for the insecurities provoked by globalization processes. This could be the main reason why 'regionalism' and 'identity' are so closely linked.

When we refer to 'identity' we must remember that 'identity' can only be made topical or actualized if 'difference' is possible. This is obviously so, because 'identity' always refers to two entities which could in principle be different, but are not. Consequently, 'identity' is only actualized by growing difference. If we see it in this way, we can understand why in late modernity, where problems of identity have become so prominent, a politics of difference, such as nationalism or regionalism, might take root.

In this sense, contemporary regionalism and nationalism are bound up with the emerging dialectics of the global and the local. While this may be an important and necessary intrinsic dilemma of late modern life, it is my concern that if the logic of regionalism and nationalism is applied to all aspects of modern life, it could become (and possibly has already become) very destructive. I will briefly point out some of these problematic implications.

The first problematic form of regionalism is certainly the process of social typification which it entails. Spatial and regional categories are used to produce stereotypes and totalizing qualifications of persons such as 'Sicilians are criminals', 'Corsicans are cunning', etc. The most crucial point here is that social or personal characteristics – positive or negative ones – are transmitted to all persons living in a certain area.

The Janus-faced, double-edged character of regionalist discourses is partly grounded in this process. Socially indifferent spatial categories – like biological ones – are used in an ideologically 'loaded' or 'charged' way for social typification. Because they are not social, they can be used in an arbitrary way. What becomes 'racist' or 'sexist' when biological categories are used for social typification becomes 'regionalist' when spatial categories are sued. All of these forms of typification undermine the rights of the subjects in modern societies and are therefore deeply anti-modern.

It is in the context of such socially typifying regionalism that political regionalism finds its preferred basis. This is because such regionalist typifications create the best conditions for instituting exclusive measures towards others, while internally the same strategy consists in the creation of identity. In the form of an excluding identity, this strategy can easily be used for political mobilization: both to create the image of an enemy and to strengthen internal solidarity. Internal differences evaporate when one emphasizes external differences.

## Globalized geographies of information and symbolization

Finally, a third research area of everyday geography could involve asking how the constitution of an actor's stock of knowledge is linked to global telecommunications (Castells 1996) and how this affects symbolization processes. This kind of informative-significative social geography is, first of all, interested in the conditions of communication, information networks and the access that particular agents have to such means of communication.

The new globalized geographies of information express the disembedding mechanisms and at the same time the main enabling tools. These tools are in McLuhan's (1995) phrasing first of all an extension of the body. But the most important characteristic is their storage and control capacity external to the body, without a definite spatial localization. Processes of information can be understood as forms of world-binding subjectively generated in relation to allocative and authoritative resources and interpretative schemes.

The core ambition of action-centred research on the geographies of information lies in the reconstruction of the access of the subjects to the flows of information and the mechanisms of control in the diffusion process. Such an analysis has to be differentiated according to the specific means and channels of communication (books, newspapers, radio, TV, data highways etc.) and their impact on the constitution of worldviews.

If the actors' stock of knowledge and information constitutes the basis of symbolic appropriation, then we have to analyse how the new geographies of information affect meaningful forms of world-binding. The analysis of the constitution of the subjects' knowledge therefore has to be linked to the constitution of the meaning-content and to symbolization processes of various areas of the everyday world. The most powerful mythological and ideological discourses are based on symbolic geographies, and are very often reproduced on the basis of reification processes as ways of naturalizing the symbolic.

The implication of the concerns outlined here is that the geographical analysis of the subjective distribution of knowledge and information – in respect to the different types of lifestyle – has to be linked to the analysis of the symbolic construction of places. Organized tradition enables a rather homogenous constitution of the meaning of places. But this is exactly what is not 'given' on the basis of late modern life forms. Hypothetically one could claim that the meaning of places depends on lifestyle and on the kind of action undertaken by a subject. Thus, two key questions for empirical research emerge: what symbolic and emotional meanings are attributed to specific places and regional representations; and what are the consequences for subsequent actions in the fields of religion, politics, economics and the appropriation of nature?

## Conclusions

In this way, an action-based social geography aims to reconstruct the everyday regionalizations of the life-world by the subjects. It critically examines the assumptive geographical representations of the world that are so often mobilized politically by regionalist and nationalist discourses. At the same time, it should become obvious that regionalism and nationalism are only two specific kinds of political attempts at regionalization. Every subject is constantly regionalizing the world through his or her actions. In this sense, any adequate geographical representation of the world has to take the subject into account: to study how subjects *live the world* and not just *in what world they live*, is one of the challenging tasks of contemporary human geography. Such a scientific geography could offer a contemporary and more appropriate view of social reality and may produce a critical account by questioning some very powerful political interpretations.

## References

Albrow, M. 1996: *The Global Age*. Cambridge: Polity Press.

Allen, J., Massey, D. and Cochrane, A. 1998: *Rethinking the Region*. London: Routledge.

Agnew, J. and Corbridge, S. 1995: *Mastering Space. Hegemony, Territory and International Political Economy*. London/New York: Routledge.

Amin, A. and Thrift, N. (eds) 1994: *Globalization, Institutions, and Regional Development in Europe*. Oxford: Oxford University Press.

Bauman, Z. 2000: *Community: Seeking Security in a Insecure World*. Cambridge.

Beck, U. 1993: *Die Erfindung des Politischen. Zu einer Theorie reflexiver Modernisierung*. Frankfurt a. M.: Suhrkamp.

Beck, U. 1997: *Was ist Globalisierung*. Frankfurt a. M.: Suhrkamp.

Beck, U. (Hrsg.) 1998: *Perspektiven der Weltgesellschhaft*. Frankfurt a. M.: Suhrkamp.

Bell, D. and Valentine, G. 1997: *Consuming Geographies. We are Where We Eat*. London: Routledge.

Braudel, F. 1990: *Sozialgeschichte des 15.–18. Jahrhunderts*, 3 vols. München: Kindler.

Carlstein, T. 1982: *Time Resources, Society and Ecology. On the Capacity for Human Interaction in Space and Time in Preindustrial Societies*. London: Allen and Unwin.

Castells, M. 1996: *The Rise of the Network Society*. Cambridge, Mass.: Blackwell.

Certeau, M. de 1988: *Die Kunst des Handelns*. Berlin: Merve-Verlag.

Chaney, D. 1996: *Lifestyles*. London: Routledge.

Crewe, L. and Lowe, M. 1996: 'United Colours? Globalisation and localisation tendencies in fashion retailing'. In Wrigley, N./Lowe, M. (eds): *Retailing, Consumption and Capital*. Harlow: Longman, 271–283.

Curry, M. R. 1996: *The Work in the World: Geographic Practice and the Written Word*. Minneapolis: University of Minnesota Press.

Entrikin, J. N. 1991: *The Betweenness of Place. Towards a Geography of Modernity*. London: Macmillan.

Featherstone, M. (ed.) 1990: *Global Culture. Nationalism, Globalisation and Modernity*. London: Sage.

Foucault, M. 1995: *Discipline and Punish: The Birth of the Prison*. New York: Vintage Books.

Fyfe, N. R. 1991: The police, space and society: the geography of policing. In *Progress in Human Geography*, vol. 15, 249–267.

Giddens, A. 1971: *Capitalism and Modern Social Theory: An Analysis of the Writings of Marx, Durkheim and Max Weber*. Cambridge, Cambridge University Press.

Giddens, A. 1981: *A Contemporary Critique of Historical Materialism, vol. 1: Power, Property and the State*. London: Macmillan.

Giddens, A. 1984: *The Constitution of Society. Outline of the Theory of Structuration*. Cambridge: Polity Press.

Giddens, A. 1985: *A Contemporary Critique of Historical Materialism, vol. 2: The Nation-State and Violence*. Cambridge: Polity Press.

Giddens, A. 1990: *Consequences of Modernity*. Stanford: Stanford University Press.

Giddens, A. 1991: *Modernity and Self-Identity. Self and Society in the Late Modern Age*. Cambridge: Polity Press.

Giddens, A. 1994: 'Living in a Post-Traditional Society'. In Beck, U., Giddens, A. and Lash, S.: *Reflexive Modernisation. Politics, Tradition and Aesthetics in the Modern Social Order*, Cambridge: Polity Press, 56–109.

Goffman, E. 1969: *The Presentation of Self in Everyday Life*. Harmondsworth: Anchor.

Goffman, E. 1971: *Relations in Public: Microstudies of the Public Order*. London: Basic Books.

Gregory, D. 1978: *Ideology, Science and Human Geography*. London: Hutchinson.

Gregory, D. 1982: *Regional Transformation and Industrial Revolution: A Geography of Yorkshire Woollen Industry*. London/Basingstoke, Macmillan.

Gregory, D. 1994: *Geographical Imaginations*. Cambridge, Mass.: Blackwell.

Guibernau, M. 1996: *Nationalisms. The Nation-State and Nationalism in the Twentieth Century*. Cambridge: Polity Press.

Habermas, J. 1998: *Die postnationale Ordnung*. Frankfurt a. M.: Suhrkamp.

Harvey, D. 1996: *Justice, Nature and the Geography of Difference*. Maden, Mass.: Blackwell.

Harvey, D. 2000: *Spaces of Hope*. Edinburgh: Edinburgh University Press.

Heidegger, M. 1986: *Sein und Zeit*. Tübingen: Max Niemeyer Verlag, (16th ed.).

Held, D., McGrew, A., Goldblatt, D. and Perraton, J. (eds) 1999: *Global Transformations. Politics, Economics and Culture*. Cambridge: Polity Press.

Jackson, P. and Thrift, N. 1995: 'Geographies of Consumption'. In Miller, D. (ed.): *Acknowledging Consumption. A Review of New Studies*. London: Routledge, 204–237.

McLuhan, M. 1995: *Die magischen Kanäle/Understanding Media*. Dresden/Basel: Verlag der Kunst.

Marx, K. 1962: *Das Kapital. Kritik der politischen Ökonomie*, vol. 1, Berlin: Dietz.

Miller, D. 1995: 'Consumption studies as the transformation of anthropology'. In Miller, D. (ed.): *Acknowledging Consumption. A Review of New Studies*. London: Routledge, 264–295.

Paasi, A. 1991: 'Deconstructing regions: Notes on the scales of spatial life'. In *Society and Space*, vol. 23, 239–256.

Philo, C. 1989: ' "Enough to drive one mad": the organization of space in 19th-century lunatic asylum'. In Wolch, J. and Dear, M. (eds): *The Power of Geography*. Boston: Unwin Hyman, 258–290.

Pickles, J. 1985: *Phenomenology, Science and Geography: Spatiality and the Human Sciences*. Cambridge: Cambridge University Press.

Sack, R. D. 1980: *Conceptions of Space in Social Thought: A Geographic Perspective*. Minneapolis: University of Minnesota Press.

Schutz, A. 1982: *Life Forms and Meaning Structures*. Trans. H. R. Wagner, London: Routledge.

Schutz, A. and Luckmann, Th. 1974: *Structures of the Life World*. Trans. R. M. Zaner and H. T. Engelhardt Jr., London: Heinemann.

Scott, A. 1999: *Regions and the World Economy: The Coming Shape of Global Production, Competition, and Political Order*. Oxford: Oxford University Press.

Shields, R. (ed.) 1992: *Lifestyle Shopping*. London: Routledge.
Sofsky, W. 1997: *Die Ordnung des Terrors: Das Konzentrationslager*. Frankfurt a. M.: Fischer.
Strohmayer, U. 1998: 'Review of Curry, M. R.: The Work in the World. Geographical Practice and the Written World'. In *Annals of the Association of American Geographers*, vol. 88, no. 1, 147–149.
Thrift, N. 1993: 'For a new regional geography'. In *Progress in Human Geography*, vol. 17, no. 1, 92–100.
Thrift, N. 1996: *Spatial Formations*. London: Sage.
Vidal de la Blache, P. H. 1903: *Tableau de la géographie de la France*. Paris: Librairie Armand Colin.
Vidal de la Blache, P. H. 1922: *Principe de la géographie humaine*. Paris: Librairie Armand Colin.
Warde, A. and Martens, L. 2000: *Eating out. Social Differentiation, Consumption and Pleasure*. Cambridge: Cambridge University Press.
Weber, M. 1968: *Economy and Society: An Outline of Interpretive Sociology*, 3 vols (ed. G. Roth and C. Wittrich, tr. E. Fischoff et al.), New York: Bedminster Press.
Werber, M. 1980: *Wirtschaft und Gesellschaft*. Tübingen: Mohr and Siebeck (5th edition).
Werlen, B. 1983: 'Methodologische Probleme handlungstheoretischer Stadtforschung'. In Lötscher, L. (ed.): *Jahrbuch der Schweizerischen Naturforschenden Gesellschaft*. Basel/ Boston/ Stuttgart, 100–109.
Werlen, B. 1987: *Gesellschaft, Handlung und Raum. Grundlagen handlungstheoretischer Sozialgeographie*. Stuttgart: Franz Steiner Verlag.
Werlen, B. 1993: *Society, Action and Space. An Alternative Human Geography*. Trans. by Gayna Walls. London: Routledge.
Werlen, B. 1995: *Sozialgeographie alltäglicher Regionalisierungen. Band 1: Zur Ontologie von Gesellschaft und Raum*. Stuttgart: Franz Steiner Verlag.
Werlen, B. 1997: *Sozialgeographie alltäglicher Regionalisierungen. Band 2: Globalisierung, Region und Regionalisierung*. Stuttgart: Franz Steiner Verlag.
Werlen, B. 2000: *Sozialgeographie. Eine Einführung*. Bern: UTB/Haupt.

# PART III
# THE SPATIAL PRODUCTION OF KNOWLEDGE

Chapter 10

# 'Local' Illustrations for 'International' Geographical Theory

Costis Hadjimichalis and Dina Vaiou

The significance of language in the construction of knowledge is a theme that has gained interest in geography, giving a different twist to the long-standing area of research on 'mapping' languages.[1] In the context of this renewed interest (for some a 'linguistic turn'), how language affects the production, consumption and circulation of geographical knowledge and how its use can imbue meaning and power to social and spatial practices is examined. Following this line of enquiry, the central role played by language in the constitution of geographical epistemologies has important implications for the kinds of discourses that develop and become dominant within the discipline. At the present time, dominant discourse undeniably originates from anglophonic literature which determines the themes and terms of geographical debate. Moreover, (American) English is the language of the single dominant country in our post-Cold War world – which intricately links theory with geopolitics.

One of the assumptions determining what 'dominant (geographical) discourse' means is that it produces 'theory', issues and concepts assumed to be universally important and relevant. Theory (or theories) thus constituted is (are) based on local knowledge which constructs and sees itself as general, without locational/spatial, or indeed other, limitations. By this token, theoretical formulations from non dominant areas (and languages) are not considered as theory, but rather as lagging examples, deviations, or local illustrations. While such attitudes would be expected from a neo-liberal perspective, it is rather puzzling for us that radical theorists succumb to the charms of comprehensive narrative/s, even when they argue strongly for the need to pay attention to difference and multiplicity.[2]

The crisis of the political left, which followed the collapse of the Eastern Block, also became evident in the production of (radical) theory in many fields in the social sciences, including geography. New concepts and categories are anxiously sought to replace 'old' formulations in many areas of research. In economic geography, Marxist-inspired radical formulations, which had gained momentum since the 1970s, came under heavy attack, as part of a grand narrative, no longer appropriate for our post-... era. In what is emerging as 'new economic geography',[3] issues like the appreciation of context for economic activity (embeddedness), its socio-institutional foundations, processes of knowledge transfer and networking, become prominent areas of research, in an attempt to 'enculturate' the economic, avoid its myopic economism and economic reductionism and expand the limits of the (sub)discipline.

The new formulations seem to follow from, and incorporate some of the arguments of, the theoretical debate on the economy-culture relation. On the cultural geography side of this debate, culture gets understood in various ways, depending on the tradition invoked (for a critical overview, see Gregson et al. 2001). But there is a distinct focus on multiculturalism and the importance of identity politics. In political geography, on the other hand, arguments about the loosening, or even disappearance, of borders (a border-less world) and the so-called 'hollowing out' of the nation state are forwarded in attempts to come to grips with the post-1989 restructuring of geopolitics. Within anglophonic geography, where these debates developed, such criticisms may be accurate, at least in part. For it is true that important aspects of what came to be included in the term 'culture' are missing from a number of accounts of economic performance in/of particular regions/areas, in a radical tradition of Althusserian structuralism,[4] while the dismantling of Cold War divisions left a real gap in understandings of geopolitical changes, at least as seen from the vantage point of the single dominant power. Many Marxists would argue that, moving the debate to the level of social formation, instead of the high abstraction of mode of production, all such 'outside' or 'non-economic' issues are – or could be – included without any need to modify theory.[5] It is surprising though how little they have indeed been included.

However, and for our purposes in this chapter, it is important to examine whether the new geography (economic, political, cultural) and the themes it introduces make space for an engagement with experiences originating in different places or whether it establishes new kinds of foreclosures. We propose to examine this by drawing examples from our own work, from a different point of departure which determines both our perspective and our focus. More specifically, we will draw (a) from research on what is comprehensively called 'the informal' and (b) from the ambiguities of our position in the Balkans and from our deep concern with the ways in which left intellectuals approached the war in Kosovo, at the time a major issue of geopolitics. Both concerns derive from our own positionality: within a different radical tradition (drawing more from Gramsci than from Althusser) and on the margins (in terms of geographical location, language and geographical debate), yet somehow touching on the dominant discourse.

## On the informal

Coming from a place where development history does not, entirely or mainly, conform to the description of the Fordist factory, the mass worker and the welfare state, over the years we have come to question the relevance of the dominant brand of (radical) geographical theory to understand and explain uneven development at various geographical scales. Analyses and explanations developed in this context project the experience of very limited areas in the world as global and undervalue the development histories of the rest, in many ways, but also by characterising as 'informal' all that is different. But what appears in the literature under the term informal, or in some cases atypical, has in our case significantly more weight than the term 'informal' allows us to believe.[6]

A point of clarification is in order here, since the term has been used, and recently re-appears, to signify different processes and phenomena. The 'informal', as it has been used since the 1980s in Southern Europe (and in our own research) is not identical with 'the household economy', 'alternative market' or 'non-market transactions' identified in the context of survival strategies in post-1989 Eastern and Central Europe (Smith 2002). Nor is its use to be confused with Third World inspired formulations about two separate economies or sectors (Roldán 1987). What we are talking about, and what has been at the core of Southern European debates, has to do with culturally embedded 'ways of doing or performing' the economic, with processes and practices which permeate the ensemble of social relations and activities and are not limited to crisis situations (for a detailed discussion, review of debates, definitions etc., see Vaiou and Hadjimichalis 1997).

Many years of research on the 'informal' gave a different turn to our understandings of urban and regional development: it has revealed to us a wealth of work and routines of everyday life associated with it which are expressed and reproduced in particular places (Vaiou and Hadjimichalis 1997). For, this umbrella term does not only include cases of excess profit-making by breaching or avoiding the law, but also conditions of hard work and strategies of survival; and inevitably draws attention to those involved in such practices, the concrete men, women, and sometimes children, local or immigrant, who contribute to the dynamism of places and bear the effects of their decline: people who are gendered and ethnicised, who work not only for the market and for pay but also within families and communities, who engage in activities rich in direct labour which is called 'informal'. This labour does not conform to descriptions of formal, full-time employment, life-long security, benefits, etc. It does not produce only surplus value, commodities and services. Through its different everyday routines, it also produces different lived spaces and times and, in this sense, it is central to understandings of their development. 'Different' here does not necessarily mean less important or lagging – definitions which can hold only from the vantage point of dominant formulations. In what follows, we briefly discuss constructions/representations of the informal focusing on three interrelated points: 'othering', a passage from derision to acknowledgement and appropriation.

## 1. 'Othering'

The brand of radical thought, which assumed a dominant position in the second half of the twentieth century, in geography as in other disciplines, developed a view of a unique global historical sequence, moving from proletarianisation and fordism to post-Fordism, which left out of theory-making all that was in some way/s different – and mainly localised in the 'South'.[7] Drawing from development experiences of the 'North', it viewed informal activities and forms of work as backward, decaying and bound to disappear. The idea of formalisation of economic and social life became one of the most widely held beliefs about spatial development and, by this token, the measuring rod for all places (Harding and Jenkins 1989). In this context and remaining only in Europe, large areas of the European South, where this progressive encroachment of formality was less obvious or slower during the post-WWII decades, became represented in radical geographical imagination as 'underdeveloped', 'backward' or 'less developed'. Such representations are also reflected in the more or less derogatory terms advanced to signal 'in-formality': hidden, black, unofficial, underground, shadow, subterranean, etc.[8] are adjectives used together with economy, sector, activity or work in order to characterise the non-conformity of such places.

In theory, the term 'informal' (or 'atypical' for that matter) determines its referent as a negation of something else – which is 'formal'. However, this is not a juxtaposition of situations to which equal importance/value is accorded. On the contrary, the term which is defined by negation is relegated to the position of 'other', inferior, less important, negatively different from an assumed norm. Thus, the places where the 'in-formal' is determinant and the subjects involved in it are seen as 'other'. And dominant geographical theory, including its radical brand, has heavily built around a binary pairing which reasserts the primacy of the norm/'formal', even in/for places where the 'in-formal' is the largest part of the economy or where it involves the majority of working population.

## 2. From derision to acknowledgement

Informal activities and forms of work started surfacing in radical geographical debate at a time which broadly coincides with significant developments in the 'North', including restructuring of large sectors of manufacturing, crisis in welfare, deregulation of formal employment (Portes et al. 1989, Chew and Denemark 1996). The argument put forward by many analysts was that the informal (or atypical) is growing in advanced and highly institutionalised economies, as full employment and the comprehensive safety net of welfare recede.[9] A change of the adjectives used is indicative of a new prevailing attitude: irregular, unregulated, undeclared, illegal are used to point to benefit fraud or 'dole cheat' committed by 'welfare spongers' or 'scroungers' (Williams and Windebank 1998; for a critique, see Gregson et al. 1999). The shifting terminology (and emphasis), however, puts in place conceptualisations which identify informal with illegal, again with negative associations which suggest a necessity for some sort of corrective measures in order to conform to the norm.

Such a turn of language is, we argue, pertinent within the 'North', but extends the idea of illegality to include activities, which in other cultural and geographical contexts might not be illegal. Acknowledgement of the informal in radical thinking thus coincides with a reassertion of 'Northern' dominance and continued construction of the 'South' as deviant. Representing the informal in these terms, ignores the voluminous research and theory-making, originating mainly from the 'South',[10] but not always available in English language. This vast literature positively acknowledges the pervasive importance of informal activities in/for Southern European societies, contributing to maintain, or even raise, living standards for large groups of people in many, and underlines their centrality in understandings of geographical development and change and in the constitution of geographical representations.

## 3. Appropriation

Informal activities and forms of work are not absent from places where the formal is dominant; indeed, their importance may have increased in recent decades, at least as far as employment is concerned (Townsend 1997). In this context, former negatively defined terms (informal, atypical, irregular and so on), often interchangeably used, assume a rather positive reference as illustrations of flexibility. 'Flexible', in new economic geography, conveys meanings to do with successful practices of capital and its responses to the 'rigidities' of Fordism. In a new round of theoretical formulations, examples from Southern European success stories are invoked, in which 'other' (informal or irregular) practices predominate – the prime example here being Third Italy.[11] Thus, practices characteristic of the deviant 'South' are selectively appropriated and inscribed in dominant radical theory.

In this process of selective appropriation, in which informal gets to be thought as identical with flexible, only certain features of Southern experiences are prioritised: small firms and firm networks, flexible adaptation to market changes, entrepreneurial spirit, innovation potential. The obvious emphasis on the firm as a unit of analysis on the one hand reasserts the primacy of the economic over those aspects which would give substance to arguments about cultural embeddedness. On the other hand it de-politicises flexibility, a term which is part of a particular political agenda which wishes to promote 'other' forms of work as a means of increasing the flexibility of capital within Europe and which, of course, is also politically contested (see, for instance, Amin and Tomaney 1995). Emphasis on the firm, then, leaves to a secondary and usually understudied plane those whose labour is appropriated by the omnipresent firm, those whose varied and insecure (informal/flexible) working patterns shape differently urban and regional development and contribute to integrate or make redundant individuals, groups and localities.

What would it mean for radical theory if it took such concerns seriously on board? An immediate response would be that there are different types of capitalism worldwide – which is not difficult to accept. Three points are worth considering here: First, such a proposition, if taken to its limits, means that the type of capitalism where 'formal' patterns predominate is one type, not *the* type, or the yardstick to which all other types are measured and found wanting. Second, theory based on that type of capitalism, on that development experience, is place-specific; therefore the claim of a globally relevant way of explaining uneven development has to be viewed with a lot of reservations. Third, and following from the previous two, when the informal is considered on the same theoretical plane, fundamental questions arise for the constitution of economic theory and its relation to geography. These three points have significant implications for theoretical formulations from a radical perspective, for the kinds of things that can or cannot be seen/discussed, the kinds of issues that can be included within the scope of debate. Beyond theory, however, they have major political implications for us as left intellectuals who prioritise the world of work rather more than the successful firm, as a guide for analysis and action.

## On the Balkans

Our second example comes from political and cultural geography, in particular from the destruction of former Yugoslavia and the latest war in Kosovo. In these furtive times, it is necessary to clarify our position from the outset. We were against war and, at the same time, we wanted to reject the 'double blackmail' which was very widespread at the time, also among left intellectuals: if you were against NATO strikes, you were for Milosevic's authoritarian regime of ethnic cleansing and if you were against Milosevic you supported the global capitalist New World Order. Instead we supported the '*Declaration of European Intellectuals for a just and permanent peace in the Balkans*' (May 1999), known also as Pierre Bourdieu's initiative. In this declaration it was argued, among other important points, that '... we have the right not to make a choice between the acquittal of Serbs' atrocities and the acquittal of NATO crimes' (for a longer elaboration of these arguments, see Hadjimichalis and Hudson forthcoming).

The war in Kosovo came at a time when radical theorists were debating, among other things, the idea of a 'borderless world', 'the hollowing out of the nation-state', 'multiculturalism' and the 'celebration of the Other' – and it has deeply divided the left in Europe. The same issues became an object of management and patronage by the military propaganda of NATO and the USA, in their attempt to legitimise what has been called, in a cynical post-modern language, 'humanitarian bombing'. What is striking for us is that years of research into the political geography of the Balkans did not seem to have provided understandings about the political interests of Great Powers. Patronage, domination and imperialism have been replaced by the fight to protect human rights and justice, thus constructing a new grand narrative in place of the heavily criticised old, modernist one.

In this context, Yves Lacoste, the French political geographer, editor of *Hérodote* and former critic of Vietnam War, not only supported NATO's intervention (*Liberation*, 20-4-1999), but suggested even map routes for land invasion from the north, via Hungary – although he later (July 1999) made a partial self-criticism. In the same vein, the majority of radical post-modernist and deconstructionist theorists, who, for many years, had criticised the one and only dominant discourse of modernism and loudly demolished the arguments of the traditional left, had not found even a word to criticise the dominant grand narrative of NATO, with the enlightened exception of Jacques Derrida (Pesmatzoglou 2000).

We argue that these, for us disturbing, attitudes are related to 'constructions'/ images of the Balkans in the dominant discourse of the West since the seventeenth century, to which the contribution of geographical thinking has been vital. Following Maria Todorova (1997), 'Balkanism' is not a case study of 'Orientalism'; it rather follows the more general and operational concept of Gramscian 'cultural hegemony'. Contrary to abstract Orientalism, which is based on a cultural antithesis, 'Balkanism' is concrete and moves around hybrid notions and cultural contradictions. Thus, we propose to examine constructions of the Balkans as articulated around three points: discovery, the power to name places and hierarchical classification.

*1. Discovery*

Balkan countries were 'discovered' as the dark background of the *Grand Tour* of young noblemen in the eighteenth and nineteenth century, who, as part of their high-class education, were visiting Italy and Greece. The focus of their visits were exclusively the ancient monuments, while all other elements of local culture, particularly those related to Islam, were seen as '... beyond the interest of civilised people'. These Grand Tours were preceded by Papal Europe which, since 1637, declared Protestants, Islam and Orthodoxy as its main enemies. It was later followed by the Habsburgs with their anti-Ottoman propaganda designed mainly for the social discipline of 'internal' populations in the Austro-Hungarian Empire. The geographical discovery of the Balkans coincided with the simultaneous invention/ construction of the area: travelers' notebooks and geographical books provided lauding descriptions of monuments and the natural environment but very devaluating ones for local population, since '… they represent a world with less civilised values than the west'.

Since World War II, the Balkans have been lost somehow behind another piece of 'Northern' (or 'Western') geopolitical imagination, the Iron Curtain. They were re-discovered only after 1989, (a) as strategic locations for the control of energy roads, which connect Europe and the Mediterranean with Russian and Iranian oil fields (Hadjimichalis 2000, van der Pijl 2001), and (b) as 'unstable neighbours … posing threats to Europe', as R. Cooper (2002) argues from his post as senior advisor to Tony Blair, legitimizing in this way the intervention of NATO in Kosovo.

## 2. The power to name places and other linguistic turns

The Balkan penninsula is perhaps the only geographical entity on mainland Europe whose name was not decided by its inhabitants. The name 'Balkan' means in Turkish a rocky, wooded mountain and was used by travelers for the Aimos mountain range, the largest in the peninsula. Its first scientific use came from the German geographer August Zeune in 1808. In the Ottoman Empire, the name *Rum-eli* was used for the peninsula, while all other ethnicities (Rumanians, Bulgarians, Greeks, Serbs, Albanians) under Ottoman occupation gave names only to local regions.

In Bulgarian, however, 'Balkani' means the area around the Aimos mountain, while its derivative *'balkandzhiya'* refers to someone with distinguished ethos, pride, courage and good will (Todorova 1997). This local definition of 'balkan' was totally reversed in the first half of the twentieth century with the introduction, by western politicians and political geographers, of the term *balkanisation* to designate the creation of small independent states in constant conflict, where violence and terror predominates. However, a more appropriate definition of the same period is the one proposed by Mower: '… small powerless states, with population and elites constantly victims of Great Powers' politics who intervene in local affairs to extend their spheres of influence' (Mower 1921, du Bois 1945).

During the war in Kosovo typically modern terms were used to describe Serbian atrocities, such as ethnic violence, genocide, new Hitler. However, a new set of post-modern terms was introduced to describe NATO attacks of more or less the same nature: surgical operations, humanitarian bombing, collateral damages, just war. The use of language for propaganda purposes is not, of course, new in situations like Kosovo. What is striking for us is the acceptance of these linguistic turns by western intellectuals, sometimes even radicals.

## 3. Hierarchical classification

The 'discovery' of 'less civilised people' and the negative connotations of the term 'balkanisation' created a representation of the region in western thought as a troublemaker, to be remembered only in periods of political uncertainty and turmoil. The construction of knowledge through such language produced meanings for the

Balkans not only of the type 'unlike us', but also as a space where all negative western characteristics of the past are dumped. Nationalisms, war, political instability, riots – which infest Western European history of the past centuries – become identified with this particular area, so that a current positive European image can be juxtaposed. What Western Europe hates most is not its image at the dawn of civilization but its image a few generations back.

Such images of the Balkans as a place of trouble and violence permeated also popular literature. For example they inspired Agatha Christie in 1925 to write a detective novel called *The Secret of Chimneys* in which the story takes place in an imagined small Balkan state where homicide prevails. Descriptions and classifications throughout the novel are apocalyptic of the deep prejudices against the Balkans: The state in question '… is a Balkan country, whose main rivers and mountains remain unknown, although it has many. Capital city, Ekaresti. Population, mainly bandits. Their hobby, regicides and revolutions …' (p. 18).

During the war in Kosovo, 74 years later, surprisingly similar descriptions were advanced, particularly in the 'balkanisation' of former Yugoslavia in the 1990s – a state seen as 'beyond the pale', not part of civilised Europe (Hadjimichalis and Hudson forthcoming). The stigma was placed on Yugoslavia and, to some extent, on the rest of non-Catholic, non-Protestant Balkan people, at a time when radical geographical discourse, in line with more general theoretical elaborations, argued for cultural difference, multiculturalism and local identity, in place of previously 'homogenizing' analyses.

For radical theory in political and cultural geography, on the one hand, the modern otherness of the area could be theoretically accepted only to the extent that it does not destabilise dominant representations; only to the extent that Balkan-ness can be contained. On the other hand, it has been appropriated by the NATO and the USA in order to legitimise military intervention and subsequent strong military presence and only to the extent that otherness remains contained under the directives of the present day Great Powers; in short, only to the extent that such otherness contributes to turn Balkan states into protectorates.

By effectively patronizing multiculturalism, difference and ethnic/local identities, wider geo-political targets have been achieved, resulting in the establishment of such protectorates, first in Bosnia, then in Kossovo and recently in FYROM. We do not argue that violence and atrocities are not part of Balkan history, past and present, and that constructed images do not hold elements of truth. Rather our point is the selectivity and the imposition of some images against others and the power asymmetries between image-makers and those represented by them.

**Summing up**

Our concerns in this paper have been largely focused on highlighting how non-dominant experiences at various geographical scales sit uneasily within a dominant discourse, these days anglophonic, which, even in its radical version/s, formulates theory and posits it as universally relevant. In economic geography, the new language, which replaces the old, totalising one, is full of references to learning firms, successful regions, embedded capital operations; uneven development, exploitation and working conditions seem to have become 'bad words' – or at least outmoded – in analyses of geographical development and change. Likewise, in political geography, globalisation, the new world order and global security seem to have replaced imperialism, aggressiveness and military power as analytical/explanatory concepts of the emerging geopolitics.

We have drawn examples from debates on the informal and from constructions/ images of the Balkans underlying the war in Kosovo, to discuss how concepts and approaches based on non-dominant experiences, if they are taken seriously, tend to destabilise dominant discourse/s and potentially open paths for more relevant radical geographical theory. In this context, we conceptualise the informal not just as a constitutive part of capitalist development, but also as a necessary element in radical politics. And in the changing geopolitics of the Balkans we identify the integration of elements of radical thinking in the totalising discourse of the aggressors, thus renewing the historical process of 'imagining' the area as dangerous and turbulent.

Forwarding these examples and arguments we do not argue for a renewed localism or empiricism. We rather envisage a debate in which radical geographical theory would actively seek to identify the terms of its international relevance, rather than assume them *a priori*, solely by virtue of its being produced in the dominant language and geographical context. During the past decade, radical geographical theory has made significant progress by incorporating feminist, cultural and postcolonial critiques. A major step forward would be to acknowledge the power relations in operation in the production of theory, relations which include questions of language, of common sense assumptions and taken-for-granted understandings about what constitutes the economy, of increasing violence and military intervention in the re-construction of geographical imaginations.

## Notes

1   See, for example, Barnes and Duncan 1992, and the Themed Section: Geographies of languages/Languages of geography, in *Social and Cultural Geography*, 2:3, Sept. 2001 (pp. 261–346).
2   Here it is probably necessary to clarify that we situate our argument in the theoretical tradition, in geography and across the social sciences, which, through many variations and internal debates since the late 1960s, has produced important understandings of the ways in which social and spatial processes and outcomes are mutually constitutive. This line of thought, known as radical geography, has contributed to 'spatialise' the social and in a sense to radicalise geography as a discipline.
3   See, for example, the relevant debate in *Antipode*, 33:2, following an intervention by Amin and Thrift (2000); also the debates in the 2nd EURS Conference.
4   They are probably not valid for other radical traditions, for example Gramscian. And this is yet another area where dominant discourse is expressed.
5   See, for example, the strong arguments put forward in the Aegean Seminars, particularly Lipietz 1996, Baltas 1996, 1999.
6   Informal refers more to the economy while atypical is used more to describe forms of employment. Interestingly, in Greek the same word is used for both terms (atypo).
7   We use here 'North' and 'South' as representations of different development trajectories and modes of thinking about them and not as a binary opposition. We are well aware of the

important differences and variation within and interpenetration between both 'North' and 'South'.

8   For a longer list see Williams, C. and Windebank, J. 1998, p. 2; also Henry 1981. The confusion of terminology, as well as the power relations involved in the corresponding representations, are a matter of recent debate; see for example Gregson et al. 1999.

9   See, however, the more nuanced analysis of Williams and Windebank 1998, which argues for coexistence and a more complex relationship between formal and informal ('multifarious "cocktails" in different places', p. 37), based on secondary material from a large number of micro-social studies in specific localities across Europe.

10  Among many: Costa Campi 1988; Mingione 1985, 1991; Paci 1992; Vaiou and Hadjimichalis 1997; Ybarra 1998.

11  It is interesting to note that Third Italy became a 'celebrity' in dominant debates after the publication by A. Bagnasco of a very short article in English (*International Journal of Urban and Regional Research*, 5:1 (1981), pp. 40–44) – which again points to the power of language.

## References

Amin, A. and Tumaney, J. (eds) 1995, *Behind the Myth of the European Union*, London: Routledge.

Amin, A. and Thrift, N. 2000, 'What kind of economic theory, for what kind of economic geography', *Antipode*, 32:1, pp. 4–9.

Baltas, A. 1996, 'Marxism's "failure": issues of meaning and responsibility', *Proceedings*, Milos Conference 'Space, Inequality and Difference. From "radical" to "cultural" formulations?', pp. 290–304.

Baltas, A. 1999, 'Wiping out history through political correctness: multiculturalism, globalisation, war', *Proceedings*, Paros Conference 'Towards a Radical Cultural Agenda for European Cities and Regions', pp. 439–446.

Barnes, T. and Duncan, J. (eds) 1992, *Writing Worlds*, London: Routledge.

Chew, S.H. and Denemark, R.A. (eds) 1996, *The Underdevelopment of Development*, London: Sage.

Cooper, R. 2002, 'The new liberal imperialism', *Observer*, 7-4-2002.

Costa Campi, M.T. 1988, 'Decentramento productivo y diffusión industriál: el modello de especialisación flexible', *Papeles de Economia Española* 35, pp. 251–276.

Du Bois W. 1945, *Color and Democracy. Colonies and Peace*, New York.

Gregson, N., Simonsen, K. and Vaiou, D. 1999, 'The meaning of work. Some arguments for the importance of culture within formulations of work in Europe', *European Urban and Regional Studies*, 6:3, pp. 197–214.

Hadjimichalis, C. 2000, 'Kossovo, 82 days of an undeclared and unjust war: a geopolitical comment', *European Urban and Regional Studies*, 7:2, pp. 175–180.

Hadjimichalis, C. and Hudson, R. (forthcoming) 'Geographical Imaginations, Identities of Neo-Imperialism and the project of European Integration', *Antipode*.

Harding, P. and Jenkins, R. 1989, *The Myth of the Hidden Economy: Towards a New Understanding of Informal Economic Activity*, Milton Keynes: The Open University Press.

Henry, S. (ed.) 1981, *Can I have it in cash? A study of informal institutions and unorthodox ways of doing things*, London: Astragal Books.

Lipietz, A. 1996, 'Cultural Geography, Political Economy and Ecology', *Proceedings*, Milos Conference 'Space, Inequality and Difference. From "radical" to "cultural" formulations?', pp. 270–283.

Mingione, E. 1985, 'Social reproduction and the surplus labour force: the case of Southern Italy', in N. Redclift and E. Mingione (eds) *Beyond Employment*, Oxford: Basil Blackwell, pp. 14–54.

Mingione, E. 1991, *Fragmented Societies*, Oxford: Basil Blackwell.

Mower P.S. 1921, *Balkanized Europe: A Study in Political Analysis*, New York.

Paci, M. 1992, *La struttura sociale italiana*, Bologna: il Mulino.

Pesmatzoglou, S. 2000, *Kosovo: The Double Offense. Supervision and Punishment*, Athens: Patakis (in Greek).

Portes, A., Castells, M. and Benton, L.A. (eds), 1989, *The Informal Economy: Studies in Advanced and Less Developed Countries*, Baltimore: Johns Hopkins University Press.

Roldán, M. 1987, 'Yet another meeting on the informal sector? Or the politics of designation and economic restructuring in a gendered world', *Proceedings*, Conference on 'The Informal Sector as an Integral Part of National the Economy', Roskilde University.

Smith, A. 2002 'Culture/economy and spaces of economic practice: positioning households in post-communism', *Transactions of the Institute of British Geographers*, 27, pp. 232–250.

Todorova, M. 1997, *Imagining the Balkans*, Oxford: Oxford University Press.

Townsend, A. 1997, *Making a Living in Europe. Human Geographies of Economic Change*, London: Routledge.

Vaiou D. and Hadjimichalis C. 1997, *... with the sewing machine in the kitchen and the Poles in the fields. Cities, regions and informal work*, Athens: Exandas (in Greek).

Van der Pijl, K. 2001, 'From Gorbachev to Kossovo: Atlantic rivalries and the incorporation of Eastern Europe', *Review of International Political Economy*, 8:2, pp. 275–310.

Williams, C. and Windebank, J. 1998, *Informal Employment in Advanced Economies. Implications for Work and Welfare*, London and New York: Routledge.

Ybarra, J.A. (ed.) 1998, *Economia sumergida: el estado de la cuestión en España*, Murcia: UGT.

# Chapter 11

# The Unhealthy and Misplaced Other

Keld Buciek

This chapter takes its point of departure in social control strategies and the power of classification. The argument is that modernity could hardly have succeeded without establishing an idea of an unhealthy Other, an atavistic being whose very existence threatens society with degeneration. Constructed during the nineteenth century, this idea has proved remarkably enduring in relations among knowledge, geography and the sociomedical labelling of those 'others' whom society regards as potentially dangerous. Consequently, the idea of purification has been a dominant metaphor for explaining social practice in Denmark as well as abroad. It is argued that inclusion and exclusion of 'the other' at home and abroad by many different means, including labelling the other as 'impure', 'unhealthy' etc., are part of the 'ordering' process of modernity as well as of modern-day immigration policies.

Very briefly, the theme is the classification of the other as unhealthy simply *because* she or he is dis- and misplaced. This kind of classification is part of social control as described by Foucault (1999, 1975), Said (1995) and Cohen (1985) among others. Such classification, this article argues, can only be combated with explicitly anti-essentialist arguments that are radically sceptical towards all categorical designations such as Us and Them, Good and Evil etc. Here I regard social control as 'the organized ways in which society responds to behaviour and people it regards as deviant, problematic, worrying, threatening, troublesome or undisirable in some way or another. This response appears under many terms: punishment, deterrence, treatment, prevention, segregation, justice, rehabilitation, reform or social defence. It is accompanied by many ideas and emotions: hatred, revenge, retaliation, disgust, compassion, salvation, benevolence or admiration. The behaviour in question is classified under many headings: crime, delinquency, deviance, immorality, perversity, wickedness, deficiency or sickness. The people to whom the response is directed, are seen variously as monsters, fools, villains, sufferers, rebels or victims' (Stanley Cohen, 1985, p. 1). The classification of 'others' is part of various political views of reality in which interesting relations among knowledge, geography and sociomedical labels are expressed.

The context of this chapter is the process by which the identity-project of Self and Other has been transformed by means of modernization into a power project involving Us and Them. Although power is at stake in every Self/Other identity project, the power project referred to here is basically a process of domination, a process of Power. This process, which entails colonization and empire-building as well as industrialization, urbanization etc, has resulted in widespread and persistent power relations associated with binary oppositions of us-them, here-there, civilization-savagery, pure-impure, enlightenment-darkness, and it is my intention to show how ambivalent notions of what are basically health-related issues, such as degeneration, *Tropenkoller* etc. have acted as lubricants of this process. As part of my research on the role of travel for our way of thinking of the world and thinking ourselves into it (see Buciek, 2001b), I have found it necessary to investigate the way western culture conceptualizes and uses notions of 'health' in discussions of space, place and mobility, viewing 'health' (and related metaphors) as part of a political vision of reality. As expressed by Said: 'If we agree that all things in history, like history itself, are made by men, then we will appreciate how possible it is for many objects or places or times to be assigned roles and given meanings that acquire objective validity only *after* the assignments are made. This is especially true of relatively uncommon things, like foreigners, mutants or "abnormal" behavior' (Said, 1995, p. 54).

Part of the discussion that this raises involves mental issues and ambivalent identity projects of various kinds. On the one hand, the aping and imitation of 'the dangerous other' is used to try to make sense of an imagined chaotic world (or of course to make money!); on the other hand the movements of this same other are restricted, among other reasons because of the fear of crime. This article addresses the central question posed over more than a hundred years by 'our' thinking: in the confrontation with 'alien' cultures (whether at home or abroad): will we be able to maintain our cultural identities, or will we degenerate in a Kurtzian manner (Griffith, 1995)? Many strategies have been used in the attempt to avoid 'going native' (apartheid is just one), but basically my argument is that the question should be deconstructed as a false question. From a post-colonial perspective health-related discussions of degeneration, *Tropenkoller*, race and climate etc. have been on the agenda for many years, most often in the context of 'culture shock', transcultural identification, decivilization, the 'Call of the Wild' or 'going native'. The tendency 'at home' has accordingly been to picture strangers only in terms of their defects, their split identities, the experience of cultural shock, the pity of homesickness, misplacement etc. (Diken, 1998). Are these elements, I am asking, still part of today's discussions of the other, of strangers in society, of immigration and refugees? And could it be that the same kind of political view of reality found in discussions during early modernity has survived the years unnoticed?

Inspired by Lévi-Strauss' dichotomized categorization of social control (absorption and disgorgement), Cohen (1985) suggests that the concepts of inclusion and exclusion systematize two rather different political visions of social control and the physiological-spatial relations between them. In Figure 11.1 I have tried to list some of the elements in these visions or semantic fields.

By exclusion Cohen means that 'deviants' are temporarily or permanently kept outside certain social boundaries or are separated and divided into their own restricted spaces. Inclusion, on the other hand, means that deviants are kept as long as possible within conventional social boundaries and institutions and are absorbed by them (ibid. p. 260). Cohen makes it clear that this should not be seen as absolute – every society mediates between these forms of social control. However, what strikes me as rather interesting is the seeming historical permanence in relations among knowledge, geography and 'sociomedical' labels in the classification of those others whom society regards as potentially dangerous.

|                              | **Inclusion**          | **Exclusion**     |
| ---------------------------- | ---------------------- | ----------------- |
| **Health and physiology metaphors** | Cannibalism    | Quarantine        |
|                              | Absorption             | Disgorgement      |
|                              | Vaccine                | Flushing          |
|                              | Therapeutic treatment  | Sluicing          |
|                              | Hygiene                | Elimination       |
|                              | Diet                   | Death-penalty     |
|                              | Sport                  | Hydros            |
| **Spatial metaphors**        | Outdoor activities     | Deportation       |
|                              | Asylum                 | Penitentiary      |
|                              | 'Bippers'              | Ghetto            |
|                              | Surveillance           | Reservation       |
|                              | Wall in                | Penal settlement  |
|                              | Imitation              | Wall              |

**Figure 11.1    A selection of metaphors used in various strategies of control**

I do not intend to discuss all aspects of the figure above, but let me just give a few preliminary examples of relations between health and space. It is well known that relations between 'boundary-transgressing' activities and the notion of 'a healthy mind in a healthy body' are cultivated in areas like sport, scouting, camping, leisure, etc. The strengthening of body and soul through exercise ('pain is good'), as a way of keeping up the health of both citizens and the nation, is pursued in areas like scouting, youth organizations, etc., all discussed in Urry (2000). Connections among tourism, health and sport have grown in recent years (a striking example is the huge combined sport-vacation-conference centre 'La Santa Sport' established in 1983 and located like a moonbase on the Canary island of Lanzarote), both building on and surpassing older forms of health resorts, sanatoria and hydros. The healing power of water, air (and fire) is associated in peculiar ways not only with individual health qualities, but also with inclusive and exclusive strategies of social control. As a very concrete example of the issue raised here, one can think of the handling of 'dangerous aliens' – that is, witches – in the past. A variety of procedures related to water, fire and exorcism were employed (the water-test, the stake etc.) to test and 'drive out' the imagined devil. As a rather more modern example to conclude this part, it has been argued by Sennett (1970) and others that the history of urban planning (including garden cities, new town projects etc.) could be read as a reaction to the dominant city metaphors of illness, cancer, slums and death, and that urban planners therefore constructed hygienic and sterile solutions (which on the other hand provoked counter-reactions – see the various essays in Westwood and Williams, 1997). Sennet express this health-space logic by stating that the essence of the mechanism of purification is the fear of loosing control (in Cohen, 1985, p. 217).

These examples in no way exhaust the subject; rather, they strengthen the focus on relations among geography, knowledge and power. It can be argued that the inclusion and exclusion of 'the other' at home and abroad by many different means, including labelling the other as 'impure', 'unhealthy' etc., are part of the 'ordering' process of modernity as well as of modern politics. In the following sections I will try to uncover elements of this 'structure of ideas'.


## Degeneration

One point of departure could be to look at the pervasive nature of ideas of atavism and degeneration in thinking about relations between 'us' and 'the other', in theories about the constitution of society, and the character of these ideas as generative metaphors. What is so fascinating about the use of these basically health-related or medical metaphors is that the majority of social scientists involved in the creation (and analysis) of the modern project seem to have rather uncritically accepted these conceptualizations. As stated by Stuart Gilman in his investigation of political theory and the works of Engels, Marx, Mill and others: 'The majority of great minds of the nineteenth century concentrated on adapting degeneracy to their own theoretical view rather than challenging it' (Gilman, 1985, p. 193). In a letter to Engels in 1866, discussing how much influence Darwin and evolutionary theory should have on their thinking, Marx wrote that 'the common Negro type is only a degeneration of a much higher one' showing that the idea of degeneration had established itself in his thinking (ibid.). However, it was Engels who grew most fond of Darwinism and degeneration theory, a tendency that is highly evident in his *The Origin of the Family, Private Property and the State* as well as in *The Part Played by Labour in the Transition of Ape to Man*. Mill's theory of 'historical deterioration' also clearly presupposes degeneracy. 'Degeneration was dominant precisely because it was so seldom questioned. It was also flexible enough to fit into most of the prevalent theoretical approaches during this era. Where it would not fit precisely, it was modified to eliminate its most negative attributes' (ibid.). The use within psychoanalysis of progression – regression terms also owes much to the concept of degeneration.

Although they were generally very optimistic about the future, in fact many writers of the late nineteenth century expressed great ambivalence towards the modern project, an ambivalence that was very much related to a widely accepted notion of the balance between civilization and barbarism, or between sanity and insanity, or between the wild and the domestic, or between the pure and the impure. In a review of Max Nordau's widely read book *Degeneration* from 1895, a reviewer stated: 'ours may be an age of progress, but it is a progress which, if left unchecked, could land us in the hospital or the lunatic asylum' (quoted in Griffith, 1995). Many cultural critics of the late nineteenth century as well as many pseudo-scientists combined what was regarded as 'knowledge' from medicine, biology, geology, Social Darwinism, craniometry, phrenology, anthropology etc. to produce theories about the imagined dark side of progress, which was then termed degeneration.

**Place and race**

The growing concern in the nineteenth century about national health issues following industrialization took as its point of departure various ideas on social biology. Ideas proposed by Darwin, Lyell and others about race and place became a platform for the discussion of national health issues, and especially the question of what happened to races when they moved out of their supposed 'designated places' in the economy of nature (Stepan, 1985). The idea of the racial type in its proper place gave the social scientist arguments about degeneration as a conceptualization of movement out of the nature-given places that different races occupied. In other words 'a race's ties to its geographical, national and social place was aboriginal and functional; it gave strength to races in their proper places. Movement out of their proper places, however, caused a "degeneration"' (Stepan, 1985, p. 99).

For the white man acting as colonizer, missionary, administrator etc. and thereby meeting 'the other', the risk of degeneration to a lower level of humankind (termed 'going native') was also there. To avoid this, a whole array of means was employed, including hygienic measurements, dress codes, rules of behaviour, drugs, baggage, diets etc. (Buciek, 2001b), not to mention apartheid strategies of many different kinds.

The idea of race very quickly became a notion that all these different social groups did not fit in society. By the mid-nineteenth century social and racial biology had established itself as a science of boundaries between groups and the degenerations that threatened humanity when these boundaries were transgressed (ibid.). Social theorists found degeneration a very compelling theme. Traces of this can be found in the works of Marx, Engels, Weber, Simmel, Pareto and many others. Degeneration is a very western reaction to urbanization, industrialization and the democratization of the political sphere (Nye, 1985, p. 52), but it has taken on a nationalist guise too. As a consequence of Germany's growing power in the nineteenth century, national health issues in France and the UK became a growing concern (Porter, 2000) because it was thought that the health/wealth of a nation and the physical and moral constitution of its people were closely related. The new social mobility and class tensions that followed industrialization and new anxieties about the 'proper place' of different class, national, and ethnic groups in society paved the way for models of the distances between social groups that were regarded as 'natural' (Stepan, p. 98): 'Racial "degeneration" became a code for other social groups whose behavior and appearance seemed sufficiently different from accepted norms as to threaten traditional social relations and the promise of "progress"' (ibid). Health issues became therefore a question of keeping social groups in touch with their natural elements at specific places. By the turn of the century those groups constructed as 'degenerates' were the urban poor, prostitutes, criminals, the insane and artists, 'whose supposed deformed skulls, protruding jaws, and low brain weights marked them as "races apart", interacting with and creating degenerate spaces near at home' (ibid). At the same time those abroad were being constructed as either children, barbarians or noble savages, depending on the situation, but with a common feature – all were seen as degenerates in the sense of not yet having reached maturity, and all were at risk of returning to an even more atavistic existence.

**Washing out atavism**

The western fear of pollution by degeneration, *Tropenkoller* etc. goes hand in hand
with the idea of exposing the body to the 'natural (and pure) environment' as a way of
strengthening body and soul. In other words, it is an attempt to answer 'the call of the
wild', by establishing situations where the individual who is 'out of place' can return
to controlled spaces and at the same time be exposed to clean air, water, earth etc. and
let the forces of nature wash out all traces of atavism and primitivism. It is remarkable
how many control strategies employed this idea, including lunatic asylums at home
and colonial strategies abroad. Since old Hippocrates, the conceptual linkage 'air –
water – place' (Porter, 2000) has been known and used. With the eighteenth century
and the spread of among other things tuberculosis concurrently with industrialization,
more and more (wealthy) people were travelling in search of balsamic air, and travel
itself became a way of establishing good health. More and more often between 1730
and 1900 combinations of travel and hydros appeared all over Europe. Hydros in
places like Baden-Baden, Vichy, Bath or Buxton became very popular, and health-
giving seaside stays, including the drinking of salty water, were also emphasized
(Porter, 2000, p. 268). Although the use of drugs and other medicaments was also
growing in importance during this period, movements within homeopathy from
c. 1800 maintained and enforced the imagined 'natural health <-> moral constitution'
paradigm. The German doctor Hahnemann's *Organon der rationellen Heilkunde*
('Guide to rational treatment') from 1810 established itself as the legacy of
Hippocrates and as the foundation of the homeopathy movement, which became
dominant in the last part of the nineteenth and the first part of the twentieth century,
gaining even wider acceptance after being recognized in 1948 by the National Health
Service in Great Britain (Porter, 2000, p. 392).

Like the homeopathy movement, the hydropathy movement, founded at the
beginning of the nineteen century by Priessnitz, emphasized purity and water. His
idea was that illness was caused by the pollution of the body by alien material, and
that drugs and medicine only contributed further to this pollution. The cure was
therefore purification through water treatment so that alien elements could by washed
out of the body. To this end, the treatment provoked sweating, followed by cold baths
and wet bandages. The first place where hydropathic principles were applied was at
Gräfenberg in Silesia around 1830, where a whole 'menu' of water treatments was
proposed to the visitor: wet stomach bandages, head-bathing where the patients had
their heads put in vessels of cold water, the douching of the genitals with ice-water,
wrapping in wet sheets etc. To this day hydropathic establishments have continued to
have an appeal for those who believe in the health-giving forces of nature, cold water
and physiological puritanism: 'no pain, no cure' (Porter, 2000, p. 393). The idea that
effort, hardship and sacrifice are or should be involved in health-related projects is an
important element in the widespread practices of hiking, walking and rambling, and is
often linked with national issues like 'Know Your Country' (for this discussion, see
Urry, 2000, pp. 52–55) and is found within sport more generally.

Not only at home, but also abroad, the idea of purification has been a dominant
metaphor for explaining social practice. One could argue that the process of
colonization has always been closely linked to ideas of cleaning dirty (primitive)
elements away. This is basically the argument in Anne McClintock's *Imperial*

*Leather.* In this marvellous analysis of the relations among race, gender and sexuality in the colonial context, soap as a commodity both at home (related to women's work, gender roles etc.) and abroad (related to the colonial subject and identification) is seen as a civilizing tool in the shaping of imperialism. Through this single household commodity the values of the new emergent middle class in the UK and other European colonial powers – monogamy and 'clean' sex, industrial capital and 'clean' money, Christianity ('being washed in the Blood of the Lamb'), class control ('cleansing the great unwashed') and the imperial civilizing mission ('washing and clothing the savage') – could all be mediated (McClintock, 1995, p. 208). In this way capitalism could claim a civilizing mission through soap advertising. 'Soap flourished not only because it created and filled a spectacular gap in the domestic market but also because, as a cheap and portable domestic commodity, it could persuasively mediate the Victorian poetics of racial hygiene and imperial progress' (ibid. p. 209). The argument can be illustrated in Figure 11.2.

**Figure 11.2    The role of soap in racism and imperialism**

Analysing soap advertising as an allegory of imperial progress as spectacle, McClintock shows how the commodity soap, as a 'magical fetish', holds out the promise of the regeneration of humankind by 'washing from the skin the very stigma of racial and class degeneration' (ibid. p. 214) and establishing a 'regime of domestic hygiene that could restore the threatened potency of the imperial body politic and the race' (ibid. p. 211). 'Victorian obsession with cotton and cleanliness was not simply a mechanical reflex of economic surplus. If imperialism garnered a bounty of cheap cotton and soap oils from coerced colonial labor, the middle class Victorian fascination with clean, white bodies and clean, white clothing stemmed not only from the rampant profiteering of the imperial economy but also from the realms of ritual and fetish' (ibid.).

*Crime, primitivism and health*

Social commentators of the time were also concerned with purifying society at home. Great attention was given to criminals, as they too were seen as products of degeneration or 'pollution from within', polluted with dispositions which, in Darwin's words, 'without any assignable cause make their appearance in families, may perhaps be *reversions to a savage state*, from which we are not removed by many generations' (my italics, quoted in Griffith, 1995).

The idea that criminals represented a particular physiological type was proposed to a wider audience by the Italian psychologist Cesare Lombroso in 1876 (he himself drew upon earlier anthropology, phrenology, Darwinism etc.). What Lombroso (and many others) did was to compare the behaviour of criminals with that of so-called primitives, because in those days too both categories were placed at the same level in the evolutionary process. Savages as well as criminals either represented those whose development had been stopped at a primitive level, or they were evolutionary 'throw-backs' (Griffith, 1995). In other words, Lombroso's *l'uomo delinquente* 'resembled a nearly morphological duplicate of the types of primitive man ... The fear of the primitive, the instinctive, and the unconscious impulses in humanity could thus be identified with a particular group' (ibid, p. 161). The fear of the primitive expressed itself both internally in relations with one's own society (with criminals, lunatics, and indeed with children), and externally in relations first with the colonial situation, towards 'the other' – here conceptualized in the notion of 'going native' (Buciek, 2001b) and secondly with 'nature' – where the notion of the 'call of the wild' summarizes the anxiety about primitivism.

Although it was born from criminology, one of the most important theories at the turn of the nineteenth century for grasping this anxiety towards primitivism (and related issues) was linked with Lombroso's concept of 'atavism'. In *Criminal Man* from 1876 Lombroso conceptualized the criminal as 'an atavistic being who reproduces in his person the ferocious instincts of primitive humanity and inferior animals' (quoted in Griffith 1995). From this controversial and widely debated elision of primitivism and criminality grew collective frameworks for the handling and

**Figure 11.3   Centrifugation machine to cure madness**
From Wimmer (n.d.).

interpreting of the social problems of industrialization, urbanization, national health etc. A common element in the handling and curing of these problems was that all that was regarded as 'impure' in society as well as in the body was thought of as something that could be hurled, flung or washed out, something that could be exorcized, moved to another place, just as the monster in Ridley Scott's *Alien* is jettisoned into space. Those regarded as insane were exposed to centrifugal forces in an attempt to fling out 'mad particles' (see Figure 11.3), those regarded as 'going native' were removed from the jungle or terminated.

### Impure aliens and ambivalent primitivism today

The above-mentioned common anxieties about relations between savagery and civilization, including the fear of the incursion of primitive man into modern culture, found expression throughout western culture in the period from 1850 on, and racism and its offspring – Nazism, Fascism etc. – carried degeneracy theory into the twentieth century, where its implications and suggested cures had their most nightmarish impact (Gilman, 1985). Does this fear of being polluted and thus of deterioration in the health of national culture lie behind modern discourses on immigration and refugees? Does it still prevail as a visible, strong force in contemporary culture? I think there are many indications that it does.

The fear of degeneration and atavism is strongly present in the works of Conrad, Melville, London, in some works by Blixen, Sandemose, J.V. Jensen; but it is also crucial to modern films like *Apocalypse Now*, *The Edge* and other 'Man against nature and other men' movies (*Extreme Limits*, *Man in the Wilderness*, *Shoot to Kill*, *Limbo*, *The Bear*, *A Cry in the Wild*, *Alaska* etc.). The fear of reversion to a primitive state is also expressed in the 2001 film *The Beach*, based on the novel by Alex Garland, where modern backpack tourism is linked with classic themes of identity, travel, Paradise/wilderness, otherness, going native, the battle between darkness and light, and the ambivalence of the western world: 'We have found the enemy and it is us'. Behind the dream of Paradise and the unspoiled tourist destination lies 'The horror! The horror!', which as in Conrad's *Heart of Darkness* basically means the fear that the traces of man's primitive past could break through the thin veneer of civilization and make any member of society degenerate, even those regarded and respected as leaders.

In recent discourses related to the wars in the Gulf and Afghanistan, as well as in discourses on immigration, with the demonization of Islam etc., generalizing arguments associating primitivism with lack of civilization, intellectual darkness and lack of democracy etc. have been proposed in the public sphere. In fact Samuel Huntington's 'clash of civilizations' seems to be nothing but a watered-down version of Nordau and others – for such criticism, see also the afterword by Said in his 1995 edition, and the Danish foreword by Botofte in the 2002 translation of *Orientalism*.

In public political debate, arguments are often drawn from the perceived deterioration of the health of the national culture. Let me briefly mention a few cases.

In 1997 the Danish right-wing political party Dansk Folkeparti tried to raise a 'concerned' debate about the number of immigrants coming to Denmark who were infected with HIV (Ritzaus Bureau, 30/11/97). A few years later the Social Democrat

Party was the platform for a proposal to isolate criminal immigrants on uninhabited islands (PolInfo, October 99). Remarks on how immigrants and refugees pollute the air on the staircases of apartment blocks (with the smell of garlic etc.) often find their way into public debate. In other words, the alien is, as a 'negative' figure, loaded with the potential for degeneration in the mainstream approach to discourse on immigration. We are dealing here with the unwelcome stranger regarded as an 'invader' of an originally healthy national body. This is also one of the arguments that Dummett (2001), in his analysis of the treatment of immigrants and refugees in Europe, spearheads. He is able to show that the arguments used in support of curtailing immigration in many European countries often make use of a perceived threat to the health of the national culture, or even claim that 'walling off' is 'good for race relations'. Recently a Danish MEP argued for war against Iraq because the 'whole Islamic culture' is 'backward, medieval, terroristic' and claimed that immigration to 'the west' would only undermine the stability of 'our society'. In his view a war would change all this to the advantage of all the 1.3 billion people living in 'Islamic cultures'!! (*Information* 27/1/03).

This is not the place to discuss the stupidity of the above arguments. What is interesting is that the structure of these kinds of arguments seems to be repeated over and over again. In other words, it seems that the receptiveness of the western audience to these themes of bestiality, impurity and primitivism has remained fairly constant over the last hundred years or more. Are we seeing a resurgence of 'theories' from the turn of the last century at the turn of this century? It is hard to say precisely – the matter should be investigated in more detail – but the point here is that a transference of individual health qualities to the collective space is clearly taking place, and this opens up the field for connecting aspects of health/purity to national culture. As stated by the Danish right-wing nationalist movement 'Fælleslisten mod Indvandring' (the front against immigration): 'The foreigners infect Danes with hideous diseases, earlier eradicated from Denmark' (www.faelleslisten.dk 24/3/04). According to 'Fælleslisten' therefore foreigners should not be given apartments. It is therefore important to ask how the idea of the unhealthy and impure other has remained so remarkably constant over so many years, or, to paraphrase Said (1995, p. 49), to ask 'what specialized skills, what imaginative pressures, what institutions and traditions, what cultural forces produce such similarity in the descriptions of the [unhealthy other] to be found in public debates, throughout so many years?'.

To answer this kind of question requires, according to Cohen (1985, p. 86), a closer look at a neglected deep structure within the system – the principle of bifurcation. The classification of the other as deviant, impure, unhealthy etc. is in no way fortuitous: 'From the foundation of the control system, a single principle has governed every form of classification, screening, selection, diagnosis, prediction, typology and policy. This is the structural principle of binary opposition: how to sort out the good from the bad, the elect from the damned, the sheep from the goats, the amenable from the non-amenable, the treatable from the non-treatable, the good risk from the bad risks, the high prediction scorers from the low prediction scorers; how to know who belongs in the deep end, who in the shallow end, who is hard and who is soft' (ibid.).

Social control strategies and the power of classification could therefore be one point of departure. There are some widespread rational but self-contradictory strategies for 'living with strangers' or coping with their ambivalence which aim at eliminating the element of pollution in the conduct of strangers. These are strategies of denial, denigration and 'ordering'. Some of the most important elements in the 'pragmatics of warfare' (Bauman's term) against the strangers seem to involve categorizing differences as absolute, creating supposedly unambivalent typologies, classifications and boundaries between 'us' and 'them', so that the grey ambivalence of the stranger is 'neutralized', made familiar as a 'known' category (Diken, 1998, p. 129), even when this means the building of walls, physically as well as in regulatory terms, as has been seen in the case of the Gypsies in Europe (Buciek, 2001a). Other strategies entail scapegoating and stigmatization. In this way difference, identity and culture are frozen and reduced to the easily recognizable category 'immigrant', which is then associated with what is problematized, unwanted, alien (Diken, 1998, p. 133).

In the view of Cohen (1985), every decision taken by the system – who is allowed to stay, who has to leave – represents and at the same time creates the basic principle of bifurcation. The specific binary assessments that dominate the system today – who is invited into the system, who is not – are part of a deep and historical structure, which is why strategies related to the handling of 'the other' throughout more than a century show so many similarities. These strategies aim at creating a '*routine world of strangers*' (Diken, 1998), that is, a world highly ordered and stripped of ambivalence and contingency. This is often directly expressed in very concrete terms (planning, architecture, social engineering etc.) and very often includes the more or less forced settlement of social groups perceived as potentially mobile 'bacteria'. 'This also explains why modernity seeking order has needed spatial forms of regulation and why those who criticized modernity, as Foucault did, did this especially by focusing on some spatial arrangements as panopticon, a modern instrument developed to bring order to the social conduct of "deviants"' (ibid, p. 130).

Before ending this discussion I should stress that we are talking about very ambivalent notions, because ideas of primitivism and the like are also used the other way round: as a way of expressing one's distance from modern society. In my view, two tendencies can be seen here. First of all, primitivism has long been a way for non-western, non-white people to express human authenticity (see Bonnett, 2000, pp. 78–118), both as a reaction to stereotyped images (as in the 'Negritude' movement), and as a genuine search for identity. Secondly, white, western people have used primitivism as an escape from an imagined cold, instrumental modernity, and by cultivating such primitivism (and very often exoticism), they have found in themselves a more 'real way of being' (ibid.). This last tendency has been accompanied by huge markets related to decoration, fashion, architecture, music and much more. Being a foreigner means first and foremost being 'dis-placed' and experiencing the disturbances that this condition causes. This is often used to approve and celebrate strangerhood as something to play upon aesthetically, rather than neutralizing it. We can think on the one hand of the tourist as the welcome stranger, on the other hand of the ever-growing 'third world industry' in contemporary societies, where cultural artefacts, modes of life, religious practices from 'primitive cultures' etc. are imitated and incorporated into modern living.

**Routes to take**

What is common to the fear of atavism, the fear of the stranger and the cultivation of primitivism is the way self-other relations are used strategically by imposing on 'the other' qualities that have more to do with the needs of 'the self' than with any acquired or inherited characteristics of the other, thereby transforming identity projects into Power projects. This transformation (and colonialism, western modernization etc.) could hardly succeed without establishing the idea of an unhealthy other, an atavistic being, from whose very existence society risks degeneration.

Where and how do we travel from here? Well, in my view the preconditions for a safe (space) odyssey include various elements. First of all, I think it would be a mistake to assume that binary oppositions between self and other are simply and only a product of western society. Ethocentricity may well be one of the most important factors through which the self is defined as a cultural construct. But what the western project has had tremendous success in establishing is the idea that, if cultural identifications are necessary in defining and maintaining what we call the 'self' in relation to the 'other', then the atavistic dissolution of a personality is a danger. Precisely this is 'The horror! The horror!' in western thought from the early European explorers to the space shuttle of today. The imagined danger of degeneration, atavism and 'going native' is that these threaten the imagined national or cultural homogeneity (or purity) of the idea of 'society'. In other word, a deconstruction of the stereotype of 'us' and 'them' (with all its related binary oppositions: here-there, clean-dirty, healthy-unhealthy, white-black, civilization-savagery, culture-nature, enlightenment-darkness etc.) must be a first step. However, to take that step requires accepting ourselves as incomplete, a reflectivity that can only be achieved the moment we recognize that the fractures between 'us' and 'them' are most likely fractures within ourselves, and that no such thing as a pure society exists.

The one-sided perception of the dangers of uprooting the stranger from his or her 'natural' origins can be balanced by looking at the stranger as a 'positive' figure. Drawing on the work of Sennett, Diken (1998) asserts that being a foreigner or a stranger can indeed be a rich experience, something that has considerable positive value. 'Displacements, of both the self and the outer context, are constructive precisely because of their disturbing or, one could say deconstructing, character; they show that the solidity of undisplaced things, as of selves which have not experienced displacement, may indeed be the greatest of illusions' (ibid, p. 134).

Of course these observations also have implications for postmodern critical ventures like the dismantling of 'society as institutions' (Urry, 2000), a project that builds heavily upon nineteenth-century thinking and on that basis creates theoretical mobilities for the twenty-first century. At the theoretical level we have to face the epistemological challenges of changing our theory of society from its primary occupation with purity, stable systems, locked-in identities etc. towards a greater emphasis on hybridity, fluid sociality, de-differentiation etc. and finding a balance between these discourses.

Whether that balance should be centred on the stranger and an ambivalent social theory where hybrid order and chaotic hybridity serve as leitmotifs, as Diken (1998) suggests, or whether other theoretical routes should be travelled, is a matter still to be decided; but at the more practical and political level there is a huge need to campaign for a more just and human treatment of aliens in Europe. The ethical and political dilemmas raised by the often unjust treatment of immigrants and refugees in several European countries seem to pose some of the most important questions that need to be dealt with at the beginning of the twenty-first century.

Dealing with these challenges no doubt means that we have to travel more, staying and learning abroad, but also that we have to invite the stranger into our society and (re)open our borders. Let me finish this essay by quoting a prominent expert on us-them relations, the former Irish president Mary Robinson, who commented on the ideal of the tolerant society in 1995 (cited in Böss, 1997, my translation): 'It is first and foremost in the recognition and acceptance of the stranger that human dignity is acknowledged. Without the other, the self would not be able to see its own humanity and the dignity connected hereto. Dignity and difference belongs to family, community and society. Both nationally and internationally ... Fellowship in difference must be the goal on every level of human life. We are growing up in a row of overlapping communities – family-related, political, cultural and religious. We mature and continue to develop as persons by combining many influences and ambitions, obligations and needs. There is a plurality in ourselves which we should learn to see and appreciate and develop. A varied society which – as it should be – is a mutual enrichment, needs varied citizens, who are able to enter into dialog with others, on the basis of the dialog in themselves.'

## Note

A first draft of this chapter was presented at '2001 – A Space Odyssey – Spatiality and Social Relations in the 21st Century', Roskilde University Field Station, Holbæk, 22–25 November 2001. It was, and is, meant as a preliminary and largely suggestive statement.

## References

Buciek, K. (2001a): 'Crossing Borders as a Way of Life – Vagabonds and Gypsies in Contemporary Social Theory'. In Bucken-Knapp, G. and Schack, M. (eds): *Borders Matter: Transboundary Regions in Contemporary Europe*. Danish Institute of Border Region Studies.

Buciek, K. (2001b): 'Stedets og rejsens ambivalens'. In Simonsen, K. (ed.): *Praksis, rum og mobilitet – Socialgeografiske bidrag*. Forlaget Samfundsvidenskab.

Böss, M. (1997): *Den irske verden: historie, kultur og identitet i det moderne samfund*. Copenhagen: Samleren.

Chamberlin, J. E. and Gilman, S. L. (1985) (eds): *Degeneration*. Columbia University Press.

Cohen, S. (1985): *Visions of Social Control – Crime, Punishment and Classification*. Cambridge: Polity Press.

Diken, B. (1998): *Strangers, Ambivalence and Social Theory*. Aldershot: Ashgate.

Dummett, M. (2001): *On Immigration and Refugees*. London: Routledge.

Foucault, M. (1977): *Overvåkning og Straff* [*The Order of Things*]. Trondheim: Gyldendal.

Foucault, M. (1999) [1966]: *Ordene og Tingene*. Viborg: Spektrum.

Gilman, S. C. (1985): 'Political Theory and Degeneration'. In Chamberlin. J. E. and Gilman, S. L. *Degeneration*. New York: Columbia University Press.

Griffith, J. W. (1995): *Joseph Conrad and the Anthropological Dilemma*. Oxford: Clarendon Press.

McClintock, A. (1995): *Imperial Leather – Race, Gender and Sexuality in the Colonial Contest*. N.Y. Routledge.

Nye, R. A. (1985): 'Sociology and Degeneration: The Irony of Progress'. In Chamberlin. J. E. and Gilman, S. L. (eds): *Degeneration*. New York: Columbia University Press.

Porter, R. (2000): *Ve og Vel – Medicinens historie fra oldtid til nutid*. Copenhagen: Forlaget Rosinante.

Said, E. W. (1995) [1978]: *Orientalism – Western Conceptions of the Orient*. London: Penguin Books.

Sennet, R. (1970): *The Uses of Disorder: Personal Identity and City Life*. New York: Alfred Knopf.

Stepan, N. (1985): 'Biology and Degeneration: Races and Proper Places'. In Chamberlin. J. E. and Gilman, S. L. (eds): *Degeneration*. New York: Columbia University Press.

Urry, J. (2000): *Sociology Beyond Societies: Mobilities for the Twenty-first Century*. London: Routledge.

Westwood, S. and Williams, J. (1997): *Imagining Cities – Scripts, Signs, Memory*. London: Routledge.

Wimmer, A. (n.y.): *Momenter af nutidens sindsygebehandling, Sct. Hans Hospital 1816–1916*, repr. in *Politikens Lægebog*, Copenhagen, 1958.

Chapter 12

# Connective Dissonance: Imaginative Geographies and the Colonial Present[1]

Derek Gregory

'Alas,' said the mouse, 'the world is growing smaller every day. At the beginning it was so big that I was afraid, I kept running and running, and I was glad when at last I saw walls far away to the right and left, but these long walls have narrowed so quickly that I am in the last chamber and there in the corner stands the last trap that I must run into.' 'You only need to change your direction,' said the cat, and ate it up.

Franz Kafka, *A Little Fable*

## Foucault's laughter

When the French philosopher Michel Foucault introduced the book that established his international reputation, *Les mots et les choses* (published in English as *The Order of Things*), he explained that it had its origins in a passage from an essay by the Argentinian novelist Juan Luis Borges. Borges had described 'a certain Chinese encyclopaedia', the *Heavenly Emporium of Benevolent Knowledge*, in whose 'remote pages' it was recorded that

[A]nimals are divided into: (a) those that belong to the emperor; (b) embalmed ones; (c) those that are trained; (d) suckling pigs; (e) mermaids; (f) fabulous ones; (g) stray dogs; (h) those that are included in this classification; (i) those that tremble as if they were mad; (j) innumerable ones; (k) those drawn with a very fine camel's-hair brush; (l) *et cetera*; (m) those that have just broken the flower vase; (n) those that at a distance resemble flies.[2]

When he read this, Foucault said that he roared with laughter, a laughter that seemed to shatter all the familiar landmarks of European thought, breaking 'all the ordered surfaces and all the planes with which we are accustomed to tame the wild profusion of existing things'. In his wonderment at this strange taxonomy, Foucault claimed to recognise the limitation of his own – 'our' own – system of thought: 'the stark impossibility of thinking *that*'.[3]

But what makes it impossible for us to think that – what lets demons and monsters loose in our own imaginary – is not so much the categories themselves. After all, the mermaids and fabulous animals are carefully distinguished from the real creatures that are variously trained, stray, and tremble. As Foucault realised, our incomprehension arises rather from the series in which they are all placed together. In short, it is not the spaces but the *spacings* that make this 'unthinkable'.

Foucault was not in the least surprised that the spacings that produced such a 'tableau of queerness' should be found in a Chinese encyclopaedia. 'In our dreamworld,' he demanded, 'is not China precisely this privileged *site* of *space?*'.

> In our traditional imagery, the Chinese culture is the most meticulous, the most rigidly ordered, the one most deaf to temporal events, most attached to the pure delineation of space; we think of it as a civilization of dikes and dams beneath the eternal face of the sky; we see it, spread and frozen, over the entire surface of a continent surrounded by walls. Even its writing does not reproduce the fugitive flight of the voice in horizontal lines; it erects the motionless and still-recognizeable images of things themselves in vertical columns. So much so that the Chinese encyclopaedia quoted by Borges, and the taxonomy it proposes, lead to a kind of thought without space, to words and categories that lack all life and place, but are rooted in a ceremonial space, overburdened with complex figures, with tangled paths, strange places, secret passages, and unexpected communications. *There would appear to be, then, at the other extremity of the earth we inhabit, a culture entirely devoted to the ordering of space, but one that does not distribute the multiplicity of existing things into any of the categories that make it possible for us to name, speak and think.*[4]

Although it would be a mistake to collapse the extraordinary range of Foucault's writings into the arc of a single project, much of his work traced just those orderings of space, at once European and modern, that appear in what he called 'the grid created by a glance, an examination, a language' – and in other registers too – which *do* 'make it possible for us to name, speak and think'. He showed with unsurpassed clarity how European modernity constructed the self – as the same, the rational, the normal – through the proliferation of spacings. But these were all spacings *within* Europe. And precisely because Foucault was so preoccupied with these interior grids – the clinic, the asylum, and the prison among them – the production of spacings that set Europe off against its exterior 'others', the very distinction between interior and exterior that initiated his journey into the order of things, was lost from view. 'The other extremity of the earth', as he called it, was literally that: extreme.

It would be perfectly possible to quarrel with Foucault's stylised characterization of China, whose cultural landscapes can be read in ways that do not confine its spaces to the bizarre and immobile geometries of French Orientalism. But that would be to miss the point. For the strange taxonomy set out in the *Heavenly Emporium of Benevolent Knowledge* was not composed by some anonymous Chinese sage. *It was invented by Borges himself.* In one sense this is unremarkable too. For, as Zhang Longxi remarks with exemplary restraint, 'What could be a better sign of the Other than a fictionalised space of China? What [could] furnish the West with a better reservoir for its dreams, fantasies and utopias?' But notice the enormous irony of it all. The nominally 'unthinkable space' that made it possible for Foucault to bring into view the modern order of things *turns out to have been thought within from within the modern too.* The joke is on Foucault. For Borges was writing neither from Europe nor from China but from 'Latin America', a topos where he was able to inscribe and to unsettle the enclosures of a quintessentially colonial modernity that Foucault was quite unable to see.[5]

## Colonial modernity: narrations and spatializations

Foucault's laughter – and the rhetorical gesture that provoked it – has become so much part of our established order of things that it is easy to forget that this order has *been* established: that it is a fabrication. This does not mean that it is simply false. On the contrary, it is validated by its own regimes of truth and it produces acutely real, visibly material consequences. Its currency – its value, transitivity, and reliability: in a word, its 'fact-ness' – is put into circulation through the double-headed coin of colonial modernity. If we remain within the usual transactions of French philosophy then one side of that coin will display the face of modernity as (for example) an optical, geometric and phallocentric space; a partitioned, hierarchical and disciplined space; or a measured, standardized and striated space. And the reverse side will exhibit modernity's other as (for example) primitive, wild and corporeal; as mysterious, capricious and excessive; or as irregular, multiple and labyrinthine. Although the coin is double-sided, however, both its faces milled by the machinations of colonial modernity, the two are not of equal value. For this is an economy of representation in which the modern is prized over – and *placed* over – the non-modern.[6]

This supplies one reason for speaking of an intrinsically *colonial* modernity. Modernity produces its other, verso to recto, as a way of at once producing and privileging itself. This is not to say that other cultures are the supine creations of the modern, but it is to acknowledge the extraordinary power and performative force of colonial modernity. Its constructions of other cultures – not only the way in which these are understood in an immediate, improvisational sense, but also the way in which more or less enduring codifications of them are produced – shape its own dispositions and deployments. These all take place within a fractured and highly uneven force-field in which other cultures entangle, engage and exert pressure. But this process of transculturation is inherently *asymmetric*, and colonial modernity's productions of the other *as* other, however much they are shaped by those various others, shape its constitution of itself in determinate and decisive ways.[7]

Here, for example, is Timothy Mitchell describing the doubled geography of a colonized, modernized Cairo at the end of the nineteenth century:

> The identity of the modern city is created by what it keeps out. Its modernity is contingent upon the exclusion of its own opposite. In order to determine itself as the place of order, reason, propriety, cleanliness, civilization and power, it must represent outside itself what is irrational, disordered, dirty, libidinous, barbarian and cowed. The city requires this 'outside' in order to present itself, in order to constitute its singular, uncorrupted identity.[8]

And here is the French anthropologist Claude Lévi-Strauss in the 1950s, gazing down on the landscape of central India from an aircraft:

> When the European looks down on this land, divided into minute lots and cultivated to the last acre, he experiences an initial feeling of familiarity. But the way the colours shade into each other, the irregular outlines of the fields and the rice-swamps which are constantly rearranged in different patterns, the blurred edges which look as if they had been roughly stitched together, all this is part of the same tapestry, but – compared to the clearly defined

forms and colours of a European landscape – *it is like a tapestry with the wrong side showing.*[9]

This extraordinary passage captures the reflexes of colonial discourse with raw precision: the European gaze on another landscape from a distance; the uncanny reflection of the European ideal ('an initial feeling of familiarity') that yields to the imperfections of the alien landscape ('irregular, blurred, rough'); and, finally, this other 'culture-nature' revealed as an image in a mirror ('a tapestry with the wrong side showing'). To Lévi-Strauss this other 'culture' and its 'nature' lack the order and clarity of his own European landscape.[10]

In his critique of Orientalism, Edward Said described this unequal process as the production of *imaginative geographies*, and Fernando Coronil connects it umbilically to what he calls 'Occidentalism'. By this he means not the ways in which other cultural formations represent 'the West', interesting and important though this is, but rather the self-constructions of 'the West' that underwrite and animate its constructions of the other.[11] It is through exactly this sort of logic that philosopher Enrique Dussel marks 1492 as the date of modernity's birth. It was only then, so he says, with Columbus's 'discovery' of America, that Europe 'was in a position to pose itself against an other' and to colonize 'an alterity [otherness] that gave back its image of itself'.[12]

But these are not merely matters of history – of a colonial past – because they reach forward into the formation of our own colonial present. In *Geography militant* Felix Driver writes about the 'worldy after-life' of colonialism, and while this does turn on 'trading in memory', as he puts it, these selective exchanges involve more than the (post)colonial cargo cult of relics and fetishes ('cultural forms') that he describes with such perspicacity. It is not just that our investments in these objects are, as he says, 'thoroughly modern': 'financial, emotional, aesthetic'.[13] For we invest in more than objects. *We also invest in a continuity of practices and dispositions.* 'Culture' and 'economy' are intimately intertwined and, as Nicholas Thomas reminds us, 'relations of cultural colonialism are no more easily shrugged off than the economic entanglements that continue to structure a deeply asymmetrical world economy'.[14]

How, then, might one understand the cultural practices that are inscribed within our contemporary tradings in memory? In one of his characteristically incisive essays on the haunting of Irish culture by its colonial past, Terry Eagleton talks about 'the terrible twins' – amnesia and nostalgia – which he describes as 'the inability to remember and the incapacity to do anything else'.[15] This provides a template within which to trace the arts of memory that play such an important part in the production of the colonial present. On the one side, then, we – a situated, privileged and prototypically white 'we', to be sure – all too readily forget the intrinsic violence of colonialism: not only the exactions, oppressions and complicities that colonialism forced upon the peoples it subjugated, but the way in which it withdrew from them the right to make their own history, ensuring that they did so emphatically not under conditions of their own choosing. We forget too, as I have indicated, the ways in which metropolitan cultures constructed its other: not only the way in which they acted as though 'the meaning they dispense[d] was purely the result of their own activity', but the way in which the predatory advances of their colonialisms appropriated other cultures and other knowledges, erasing the process of transculturation altogether.[16] On the other side, contemporary metropolitan cultures are disfigured by a yearning for the aggrandizing swagger of colonialism itself: for its privileges and its powers. Its exercise may have been shot through with anxiety, even guilt; its codes may (on occasion) have been transgressed, even set aside. But the triumphal *show* of colonialism – its 'ornamentalism', as David Cannadine calls it, and its effortless, ethnocentric assumption of Might and Right[17] – are visibly and aggressively abroad in our own world. Ironically, there is also a plangent yearning for the cultures that the impress of colonial modernity virtually destroyed. Art, design, fashion, film, literature, music, travel: all are marked by a nostalgic – and thoroughly commodified – longing for what Graham Huggan calls the (post)colonial exotic.[18]

The constellation formed by these conjunctions is tense, agonistic – a sort of present tense – and this speaks directly to the slippery spatiality of the colonial present. These are not only matters of time and history, because the 'arts of memory', as Francis Yates and Walter Benjamin have shown in such strikingly different ways, of forgetting and remembering, also turn on space and geography: on the performance of imaginative geographies.

## September 11 and the colonial present

I want to sharpen these claims through a consideration of some of the imaginative geographies that articulated the colonial present in the aftermath of *al-Qaeda*'s attacks on metropolitan America on 11 September 2001. In the days and weeks that followed September 11, the compound series of events knotted around the attacks on the Pentagon in Washington and the World Trade Center in New York City, the asymmetry that underwrites the colonial production of imaginative geographies was endlessly elaborated through the repetition of a single question. 'Who hates America?' asked Michael Binyon in the London *Times* two days after the attacks. On 20 September President George W. Bush appeared before a joint session of Congress and acknowledged the same, central question that had come to dominate public discussion: 'Americans are asking, "Why do they hate us?"'.[19]

As Roxanne Euben remarked, however, the question is itself a kind of answer, because it reveals 'a privilege of power too often unseen: the luxury of not having *had* to know, a parochialism and insularity that those on the margins can neither enjoy nor afford'.[20] It is this asymmetry – accepting the privilege of contemplating 'the other' without acknowledging the gaze in return, what novelist John Wideman calls 'dismissing the possibility that the native can look back at you as you are looking at him' – that marks this as a colonial gesture of extraordinary contemporary resonance. As Wideman goes on to say, 'The destruction of the World Trade Center was a criminal act, the loss of life an unforgivable consequence, but it would be a crime of another order, with an even greater destructive potential, to allow the evocation of the word "terror" to descend like a veil over the event, to rob us of the opportunity to see ourselves as others see us'.[21]

And yet for the most part American public culture was constructed through a one-way mirror. The metropolis exercised its customary privilege to inspect the rest of the world. On 15 October 2001 *Newsweek* produced a thematic issue organized around the same question: 'Why do they hate us?'. The answer, significantly, was to be found among 'them' not among 'us': not in the foreign policy adventures of the USA, for example, but in the chronic failure of Islamic societies to come to terms with the modern. The author of the title essay, Fareed Zakaria, explained that for bin Laden and his followers 'this is a holy war between Islam and the Western world'. Zakaria continued like this:

> Most Muslims disagree. Every Islamic country in the world has condemned the attacks … But bin Laden and his followers are not an isolated cult or demented loners … They come out of a culture that reinforces their hostility, distrust and hatred of the West – and of America in particular. This culture does not condone terrorism but fuels the fanaticism that is at its heart.[22]

This is an instructive passage because its double movement was repeated again and again over the next weeks and months. In the opening sentences Zakaria separates bin Laden and his followers from 'most Muslims', who disagreed with their distortions of Islam and condemned the terrorist attacks carried out in its name. But in the very next sentences the partition is removed. Even those Muslims who disagreed with bin Laden and condemned what happened on September 11 are incarcerated with the terrorists in a monolithic super-organic 'culture' that serves only to reinforce 'hostility, distrust and hatred of the West' and to 'fuel fanaticism'. The culture of 'Islam' – in the singular – is made to absorb everything, as though it were a black hole from whose force-field no particle of life can escape. This is a culture not only alien in its details – its doctrines and observances – but in its very essence.[23] This culture ('their' culture) is closed and stultifying, monolithic and unchanging, whereas 'our' culture (it goes without saying) is open and inventive, plural and dynamic. And yet, as Mahmood Mamdani asked in the wake of September 11,

> Is our world really divided into two, so that one part makes culture and the other is a prisoner of culture? Are there really two meanings of culture? Does culture stand for creativity, for what being human is all about, in one part of the world? But in the other part of the world, it stands for habit, for some kind of instinctive activity, whose rules are inscribed in early founding texts, usually religious, and museumized in early artefacts?[24]

If our answer is yes, then we are saying that 'their' actions do not derive from any concrete historical experience of oppression or injustice, or from the imaginative, improvizational practices through which we ceaselessly elaborate our world. 'Their' actions are simply dictated by the very nature of 'their' culture. When Zakaria rephrased his original question it was only to make the same distinction in a different way. 'What has gone wrong in the world of Islam?' he asked. To see that this *is* the same distinction one only has to reverse the question: 'What has gone wrong in America?'. From Zakaria's point of view, the only answer possible would be 'the terrorist attacks on New York City and Washington'. Not only is culture partitioned; so too is causality. 'America' is constructed as the normal – because it is assumed to be the universal – and so any 'attack on America' can only have arisen from the pathologies that are supposed to inhere within 'the world of Islam'.

In a subsequent essay Zakaria made his base assumption explicit: 'America remains the universal nation, the country people across the world believe should speak for universal values.'[25] This was not an exceptional view, and Veena Das drew attention to the way in which, even more so after September 11, America constructed itself as 'the privileged site of universal values'.

> It is from this perspective that one can speculate why the talk is not of the many terrorisms with which several countries have lived now for more than thirty years, but with one grand terrorism – Islamic terrorism. In the same vein the world is said to have changed after September 11. What could this mean except that while terrorist forms of warfare in other spaces in Africa, Asia or the Middle East were against forms of particularism, the attack on America is seen as an attack on humanity itself ... It is thus the reconfiguration of terrorism as a grand single global force – Islamic terrorism – that simultaneously cancels out other forms of terrorism and creates the enemy as a totality that has to be vanquished in the interests of a universalism that is embodied in the American nation.[26]

In case my comments are misunderstood, I should say that there are indeed substantial criticisms to be made of repressive state policies, human rights violations and non-democratic political cultures in much of the Arab world. But many (most) of those regimes were set up or propped up by Britain, France and the United States. And it is simply wrong to exempt America from criticism, and to represent its star-spangled banner as a universal standard whose elevation has been inevitable, ineluctable: in a word, simply 'natural'.[27] As Arundhati Roy asked: 'Could it be that the stygian anger that led to the attacks has its root not in American freedom and democracy, but in the US government's record of commitment and support to exactly the opposite things – to military and economic terrorism, insurgency, military dictatorship, religious bigotry and unimaginable genocide?'[28] This was a difficult, even dangerous question to pose immediately after September 11 when attempts at explanation risked being condemned as exoneration.[29] And yet, precisely because these are simultaneous equations in what Roy called 'the fastidious algebra of infinite justice', Bush's original question – 'Why do they hate us?' – invited its symmetrical interrogatory. When ordinary people in Iraq or Afghanistan cowered under US bombardments, they surely asked themselves, 'Why do they hate us?'.[30]

It is not part of my purpose to adjudicate between these questions and counter-questions because it is the dichotomy reproduced through them that I want to contest. As Said argues, 'to build a conceptual framework around the notion of us-versus-them is in effect to pretend that the principal consideration is epistemological and natural – our civilization is known and accepted, theirs is different and strange – whereas in fact the framework separating us from them is belligerent, constructed and situational'.[31]

## Performance, space and the postcolonial

In Said's original discussion, which was informed in all sorts of ways by the Foucault of *Discipline and Punish* rather than *The Order of Things*, imaginative geographies fold distance into difference through a series of spatializations. They multiply partitions and enclosures that demarcate 'the same' from 'the other', at once constructing and calibrating a gap between the two by 'designating in one's mind a familiar space which is "ours" and an unfamiliar space beyond "ours" which is "theirs"'. Said's primary concern was with the ways in which European and American imaginative geographies of 'the Orient' combine over time to produce an internally structured archive in which things come to be seen as neither completely novel nor thoroughly familiar. Instead, a median category emerges that 'allows one to see new things, things seen for the first time, as versions of a previously known thing'.[32]

This Protean power of Orientalism is immensely important – all the more so now that Orientalism is abroad again, revivified and hideously emboldened – because the citationary structure that is authorized by these accretions is also in some substantial sense *performative*. In other words, its categories, codes and conventions shape the practices of those who draw upon it, actively constituting its object (for example, 'the Orient') in such a way that this structure is as much a *repertoire* as it is an archive. This matters for two reasons. In the first place, as the repertory figure implies, imaginative geographies are not only accumulations of time, sedimentations of successive histories; they are also *performances of space*.[33] In the second place, performances may be scripted (they usually are) but this does not make their outcomes fully determined; rather, performance creates a space in which it is possible for 'newness' to enter the world. This space of potential is always conditional, always precarious, but every performance of the colonial present carries within it the possibility of undoing its enclosures and approaching closer to the horizon of the *post*colonial.[34]

Imaginative geographies, then, simultaneously conjure up and hold at bay the strange, the unnatural, the monstrous. 'Given their monstrosity,' Zygmunt Bauman declares, 'one cannot but thank God for making them what they are – the *far away* locals, and pray that they stay that way'. But distance is never an absolute, fixed and frozen, and within the colonial present, like the colonial past, the power to transform distance – like the power to represent others *as* other – is typically arrogated by metropolitan cultures. Bauman distinguishes between 'residents of the first world' – 'tourists', he calls them – who live pre-eminently in time, who can span every distance with effortless ease, and who move 'because they want to', and 'residents of the second world' – 'vagabonds' – who live pre-eminently in space, 'heavy, resilient, untouchable', and who travel surreptitiously and often illegally 'because they have no other bearable choice'.[35]

Of course, this is an unstable distinction. Bauman's 'tourists' depend on the 'vagabonds' in all sorts of ways, not least on their cheap labour as unlicensed, undocumented migrants. If this is a caricature, however, it is at least a recognizable one. It suggests that part of the shock of September 11 was brought about by its abrupt reversal of metropolitan privilege. On that bright morning, distance was spectacularly compressed and liquid modernity turned into fire. The horror, says Bauman, 'brought the untouchable within touch, the invisible within sight, the distant within the neighbourhood.'[36] Here is American novelist Don de Lillo:

> Our world, parts of our world, have crumbled into theirs, which means we are living in a place of danger and rage ... The terrorists of September 11 want to bring back the past ... The future has yielded, for now, to medieval experience, to the old slow furies of cut-throat religion ... Now a small group of men have literally altered our skyline. We have fallen back

in time and space ... There is a sense of compression, plans made hurriedly, time forced and distorted.[37]

Time and space crumbled, collapsed, compressed: not by 'us' but by 'them'; not on our terms but on 'theirs'.

'They' were not Bauman's 'vagabonds' – those responsible for the attacks on New York City and Washington were hardly the wretched of the earth – but they did come from a world away. And that world had been made strange – alien, monstrous – long before transnational terrorism erupted in the heart of metropolitan America. 'Terror has collapsed distance', wrote Michael Ignatieff, 'and with this collapse has come a sharpened American focus on bringing order to the frontier zones.'[38] Similarly, for John Urry September 11 was a dramatic rupture in which the world's 'safe zones' and 'wild zones' – 'civilization' and 'barbarism' – 'collided in the sky above New York'. But he also saw it as an event that brought into view a vast, almost subterranean movement whose time-space compressions at once preceded the attacks and extended beyond them:

> The flows from the wild zones of people, risks, substances, images, Kalshnikovs ... increasingly slip under, over and through the safe gates, suddenly and chaotically eliminating the invisibilities that had kept the zones apart. Through money laundering, the drug trade, urban crime, asylum-seeking, arms trading, people smuggling, slave trading and urban terrorism, the spaces of the wild and the safe are chaotically juxtaposed, time and space is being 'curved' into new complex configurations.[39]

What is missing from all these characterizations, however, is the recognition that the power to compress distance is also the power to *expand* distance. The unidirectional logic of what David Harvey calls 'time-space compression' requires distance to contract under the compulsions of global capitalism to reduce circulation time. This has its own uneven, inconstant geography, which short-circuits across economic, political, military and cultural registers, but images of 'the global village', 'the shrinking world' and 'the space of flows' have become so powerful and pervasive that, as Cindi Katz observes, they have obscured the ways in which distance also expands.[40] This happens not only within a relative space punctuated by the differential geographies of time-space compression – Katz's (sharp) point – *but also through the proliferating partitions of colonial modernity that set 'us' apart from 'them'.*[41]

These torsions – time-space compression and time-space expansion – work through two contradictory logics. Michael Hardt and Antonio Negri offer a first approximation in their otherwise contentious account of the formation of the contemporary constellation of power which they call 'Empire':

> Imperialism is a machine of global striation, channelling, coding and territorializing the flows of capital, blocking certain flows and facilitating others. The [capitalist] world market, in contrast, requires a smooth space of uncoded and deterritorialized flows.[42]

If global capitalism is aggressively *de-territorializing*, moving ever outwards in a process of ceaseless expansion and furiously tearing down barriers to capital accumulation, then colonial modernity is intrinsically *territorializing*, forever installing partitions between 'us' and 'them'. I describe this as a first approximation because Hardt and Negri go on to suggest that 'Empire' is busily resolving this contradiction by dissolving the distinction between 'outside' and 'inside'. 'There is no more Outside,' they write. 'The modern dialectic of inside and outside has been replaced by a play of degrees and intensities, of *hybridities* and artificiality.' Similarly, Bauman: 'September 11 made it clear that *'Il n'y a pas du "dehors"* any more.'[43] But this is to claim too much (and to accept too many of postcolonialism's preoccupations).[44] For over two hundred years, as Das cogently reminds us, 'the distinction between an "inside" in which values of democracy and freedom were propagated and an "outside" which was not ready for such values and hence had to be subjugated by violence in order to be reformed has marked the rhetoric and practice of colonialism and its deep connections with Western democracies'.[45] Das had good reason to say this in the aftermath of September 11, and the subsequent unfolding of 'the war on terror' in Afghanistan and its violent extensions into Palestine have demonstrated not the slackening but the tightening of this colonial spacing.[46]

In fact, colonialism's promise of modernity has always been deferred – always skewed by the boundary between 'us' and 'them' – and although that partition is routinely crossed, even transgressed, the dismal fact is that no colonial anxiety, no colonial guilt has ever erased it altogether. If this is still the primary meridian of imaginative geography, however, it is no simple geometry. It is, rather, a topology whose seams are contorted and twisted into new, ever more wrenching constellations. Through these torsions the divide may be annulled in some registers (in these ways, you may be modern: like 'us') while it is simultaneously reaffirmed in others (in these ways you will never be modern: always irredeemably 'other'). The effect of these differential torsions is exceptionally significant. For the modern world is indeed marked by – and produced through – spacings of connection, worked by transnational capital circuits and commodity chains, by global flows of information and images, and by geopolitical alignments and military dispositions. But the modern world is also marked by – and produced through – spacings of disjuncture between 'us' and 'them', between 'the same' and 'the other'. Imaginative geographies are thus doubled spaces of articulation. Their inconstant topologies are mappings of *connective dissonance* in which connections are elaborated in some registers even as they are disavowed in others.

So, for example, on September 11 America was attacked by *al-Qaeda* and in response attacked *al-Qaeda* bases and the Taliban in Afghanistan. The cause-and-effect connections seem clear. And yet the chains linking these translocal compressions to the guerilla war fought by the *mujaheddin* against the Soviet occupation of Afghanistan, with support from the USA, Pakistan and Saudi Arabia, and to America's aggressive pursuit of its own geopolitical and geoeconomic interests in the region, have been all but sundered within the American public imaginary. Or again, Israel was attacked by Palestinian suicide bombers, and – under the cover of America's 'war on terrorism' – launched massive retaliations against the occupied territories. Again, the cause-and-effect connections seem clear. But the roots of these murderous attacks, and the militarization of the *al-Aqsa* Intifada more generally, in the grinding, dehumanizing violence of the illegal Israeli military occupation have been virtually torn up.[47] Elsewhere I have described these wrenching articulations as 'gravity's rainbows'. In Thomas Pynchon's extraordinary novel of (more or less) that name, the arc of the V-2 rockets launched from occupied Europe against Britain in the dying days of World War II is described as 'a screaming coming across the sky'. But this arc traces a physical connection that is *also* a moral disconnection: an unwillingness or an inability to hear an altogether different sound: the screaming of its victims on the ground. So, too, do imaginative geographies map the twists and turns of engagement and estrangement.

## The colonial present and cultures of travel

Mappings like these articulate contemporary cultures of travel. Bauman's 'tourists' probably know this without being told; at least those accustomed to move from one singular, out-of-the-ordinary, even exotic site/sight to another, gazing upon the other but always able to withdraw to the security of the familiar, the known, know this. They move in the folds between compression and expansion, and their performances are profoundly colonial gestures.[48] In fact, cultures of travel are some of the most commonplace means through which colonialism is insistently abroad in our own present. If, as Urry once remarked, it is now the case that in the first world 'people are much of the time tourists, whether they like it or not', it is also the case that they – we – are intimately implicated in the perpetuation of the colonial present.[49] Ironically one of the short-term consequences of September 11 was to contract the space of American tourism as flights were cancelled and aircraft flew half-empty. Two weeks later Bush told enthusiastic airline employees at Chicago's O'Hare airport that 'one of the great goals of this war is to tell the traveling public: Get on board'.[50] This must count as one of the most bizarre reasons for waging war in human history, and yet it also speaks a powerful truth. Modern metropolitan cultures privilege their own mobility.

'Privilege' has to be understood literally; there are other cultures of travel within which movement is a burden, an imposition, even a tragedy. What, then, of Bauman's 'vagabonds?'. Three weeks after September 11 the metropolis reasserted its customary powers and privileges as military action was launched against Afghanistan, and thousands of refugees were displaced by these pulverizing time-space compressions. Many of them were trapped at borders – not only at Afghanistan's borders but also at other borders around the world. Here is Gary Younge on their experience almost-in Britain:

> [S]hould those whom we seek to protect [by our international military actions] arrive on our shores, all apparent concern evaporates in a haze of xenophobic bellicosity. Whatever compassion may have been expressed previously is confiscated at the border. As soon as they touch foot on British soil they go from being a cause to be championed to a problem to be dealt with. We may flout international law abroad, but God forbid any one who should breach immigration law here ... We love them so we bomb them; we loathe them so we deport them.[51]

Thousands of displaced people, refugees and asylum-seekers found that, in the very eye of these wrenching time-space compressions, time and space had dramatically expanded for them. 'The globe shrinks for those who own it', Homi Bhabha once remarked, but 'for the displaced or the dispossessed, the migrant or refugee, no distance is more awesome than the few feet across borders or frontiers'.[52] It should come as no surprise, therefore, to find that Suvendri Perera can draw a close parallel between the extra-territoriality of Camp X-Ray at Guantanamo Bay, where hundreds of so-called 'unlawful combatants' are detained not only outside the territorial United States but outside the protections of international law, and Woomera Detention Centre, where asylum-seekers are held 500 km from Adelaide: 'a place that is, and yet is not, Australia'. 'Razor wire and metal fencing mark out the camp as a space of exception. Five layers of wire protect the threshold between Australia and its other, not-Australia.'[53] It is appropriate to remind ourselves that camps like these have their origins in European colonial wars of the late nineteenth century. In conditions like these, now as then, the 'external' and the 'internal' are articulated not to erase the 'outside' but to produce it *as the serial spacing of the exception*, forever inscribing exclusion through inclusion.

Giorgio Agamben describes these spacings as zones of indistinction, spacings of the exception, and as they multiply around the world, we are forced to recognize that 'third spaces' or 'paradoxical spaces' are not always the emancipatory formations that some contemporary writers have taken them to be.[54] After September 11 Agamben worried that the 'war on terrorism' would be invoked so routinely that the exception would become the rule, that the law would be forever suspending itself.[55] His fears were well-founded. Less than a year later the US Director of Human Rights Watch reported that 'the US government has failed to uphold the very values that President Bush declared were under attack on September 11'. The Bush administration attempted to curtail democratic freedoms in at least three arenas, by circumventing federal and international law, by suppressing public information (even the conservative Cato Institute objected to the proliferation of 'secretive subpoenas, secretive arrests, secretive trials, and secretive deportations'), and by discriminating against visible minorities.[56] This series of exceptions was consistent with – and legitimised by – the imaginative geographies of 'civilization' and 'barbarism' that were mobilized by the White House and which articulated 'a constant and mutual production of the civilized and the savage *throughout* the social circuitry.' These spacings folded 'inwards' as well as 'outwards', therefore, to produce 'a constant scrutiny of those who bear the sign of "dormant" terrorist and [to] activate a policing of points of vulnerability against an enemy who inheres within the space of the US'.[57]

To be sure, this process has not been confined to the United States. When Kanishka Jasuriya warned against the creeping internationalization of the state of exception – already in train before September 11 – he drew special attention to the application of these strategies of risk management and profiling to 'target populations'.[58] 'Arabs' and 'Muslims' have been made desperately vulnerable by these identifications. One fifth of Israel's own population has increasingly been excluded from politically qualified life; Israeli Arabs have been demonised not only as a fifth column but as a 'cancerous growth', and in the occupied territories – for which Israel takes little or no civil responsibility – Sharon's chief of staff has chillingly promised the Palestinians 'chemotherapy'.[59] In Europe and Australia there has been a frenzied assault on immigrants, refugees and asylum-seekers. At the beginning of *The Clash of Civilizations*, which was so often invoked in the wake of September 11, Samuel

Huntington cites a passage from novelist Michael Dibdin's *Dead Lagoon* that speaks directly to these issues: 'There can be no true friends without true enemies. Unless we hate what we are not, we cannot love what we are.' Huntington described this as 'an unfortunate truth', as though it were somehow inscribed in the nature of things. But in the novel these sentences come from a speech given by Ferdinando Del Maschio to drive home the cause of self-determination for the once serene republic of Venice and the need to preserve and purify its Venetian culture. At the end of the novel Del Maschio rounds on Dibdin's detective, Aurelio Zen, and tells him:

> Sooner or later you're going to have to choose ... The new Europe will be no place for rootless drifters and cosmopolitans with no sense of belonging. It will be full of frontiers, both physical and ideological, and they will be rigorously patrolled. You will have to be able to produce your papers or suffer the consequences.[60]

Dibdin is surely saying that this *is* a choice and not, as Huntington seems to think, an irrefusable give, so that the construction of an archipelago of inclusion and exclusion – an architecture of enmity – cannot be attributed solely to the threat posed by external or internal 'others'. In constructing multiple others *as* 'other', and in assenting to these constructions and impositions, we not only do this to others: *we do this to ourselves*. We all become the subjects *and the objects* of the 'securitisation of civil society'. This is as ugly as it sounds – it means taking the 'civil' *out* of 'society' – and as its partitions proliferate internally and externally, inscribed through and legitimised by the 'war on terrorism', so colonialism is surreptitiously repatriated and rehabilitated. The camp is confirmed as the *nomos* of colonial modernity.

And yet, if we can understand the multiple ways in which difference is folded into distance, and the complex figurations through which time and space are folded into such tense constellations, we might see that what Michael Ignatieff once called 'distant strangers' are not so distant after all – and not so strange either.[61] We need to explore other spatializations and other topologies, and to turn our imaginative geographies into geographical imaginations that can enlarge and enhance our sense of the world and enable us to situate ourselves within it with care, concern and humility.[62] This is not a call for an empty relativism; there will still be disagreements, conflicts and even enemies. But in order to conduct ourselves properly, decently, we need to set ourselves against the unbridled arrogance that assumes that we have the monopoly of Truth and that the world is necessarily ordered by – and around – Us. If we can do this, then we might also see that the most enduring memorial to the thousands who were murdered in the catastrophic destruction of the World Trade Center and parts of the Pentagon on September 11 – and to the thousands who have been murdered in Afghanistan, Israel and Palestine – would be the destruction of the imaginative geographies that produced and have been sustained by these dreadful events. We might then begin to approach the postcolonial.

## Notes

1   This essay is drawn from *The Colonial Present* (Oxford: Blackwell, 2004). I am grateful to Kirsten Simonsen for the opportunity to present a preliminary version of my argument in

Denmark, and to the participants in the 'Space Odyssey' for their help and encouragement in my journey.

2   Jorge Luis Borges, 'John Wilkins' Analytical Language', in his *Selected Non-fictions* (New York: Viking, 1999) pp. 229–232; quotation from p. 231. The essay was first published in Spanish in 1942.

3   Michel Foucault, *The Order of Things: An Archaeology of the Human Sciences* (London: Tavistock, 1970) p. xv; first published in French as *Les mots et les choses* (Paris: Gallimard, 1966).

4   Foucault, *The Order of Things op. cit.*, p. xix; emphasis added. I have taken 'tableau of queerness' from Edward Said, *Orientalism* (London: Penguin, 1995; first published in 1978) p. 103.

5   Zhang Longxi, *Mighty Opposites: From Dichotomies to Differences in the Comparative Study of China* (Stanford CA: Stanford University Press, 1998) pp. 21–2. Foucault does of course concede that his description of China as a 'privileged site of space' belongs to 'our dreamworld': but he does not pursue those spectral footprints. In Borges's original essay, in contrast, the 'Chinese encyclopaedia' was not summoned up to represent a system of thought alien to European culture, since its (imaginary) 'absurdities' were used to parallel the (real) 'ambiguities, redundancies and deficiencies' in the attempt made by the English cleric John Wilkins to develop a universal language in his *Essay towards a Real Character and a Philosophy of Language* (1688). For a discussion of the hybrid spaces between magical realism more generally and the critique of colonialism, and of the vexed politics to which they give rise, see Stephen Slemon, 'Magic realism as post-colonial discourse', *Canadian Literature* 116 (1998) pp. 9–24.

6   It is for this reason that J.K. Gibson-Graham objects to these characterizations as so many versions of a 'rape-script' that works to normalize an act of non-reciprocal penetration. Through its developing sequence of steps and signals, all non-modern, non-capitalist forms are rendered as sites of a lack, always and everywhere targets of invasion and submission; they are constructed as inevitably damaged, fallen, violated, waiting to be mastered by the transcendentally powerful spatializations of an advancing capitalist, colonizing modernity. See J.K. Gibson-Graham: *The End of Capitalism (as we knew it): A Feminist Critique of Political Economy* (Oxford: Blackwell, 1996).

7   The concept of transculturation was developed by the Cuban ethnographer Fernando Ortiz, who juxtaposed the term to 'acculturation': whereas acculturation implies the adjustment of a subordinate culture to the impositions and exactions of a dominant culture, transculturation implies a dynamic relation of combination and contradiction. For a critical discussion of the genealogy of the concept, see John Beverley, *Subalternity and Representation: Arguments in Cultural Theory* (Durham NC: Duke University Press, 1999) pp. 43–7, and for its deployment see Mary Louise Pratt, *Imperial Eyes: Travel Writing and Transculturation* (London: Routledge, 1992).

8   Timothy Mitchell, *Colonising Egypt* (Cambridge: Cambridge University Press, 1988) p. 165; I complicate this duality in 'Rumours of Cairo', in my *Dancing on the Pyramids* (Minneapolis: Minnesota University Press, forthcoming).

9   Claude Lévi-Strauss, *Tristes tropiques* (London: Penguin, 1992; original French publication 1955) pp. 132–3; emphasis added.

10  For a detailed discussion, see my '(Post)colonialism and productions of nature', in Noel Castree and Bruce Braun (eds) *Social Nature: Theory, Practice, Politics* (Oxford: Blackwell, 2001) pp. 84–111.

11  Said, Orientalism *op. cit*; Fernando Coronil, 'Beyond Occidentalism: toward non-imperial geohistorical categories', *Cultural Anthropology* 11 (1996) pp. 51–87. Said thought it unlikely that anyone would imagine 'a field symmetrical to [Orientalism] called Occidentalism' because the imaginative geographies produced by other cultures are not

bound in to a system of power-knowledge comparable to the tensile strength and span of European and American colonialism and imperialism. In his view (and mine) it would be wrong to suppose that reversing the terms equalizes them, and it is this fundamental *asymmetry* that Coronil seeks to redeem through his (contrary) reading of 'Occidentalism'.

12  Enrique Dussel, 'Eurocentrism and modernity', *Boundary 2* 20.3 (1993) pp. 65–76; see also his *The Invention of the Americas: Eclipse of 'The Other' and the Myth of Modernity* (New York: Continuum, 1995). Dussel notes that '1492' is itself doubled, marking not only Columbus's voyage to the Americas but also the Christian *Reconquista* that snuffed out Islamic rule in Andalusia. He argues that the sectarian violence unleashed in the closing stages of the *Reconquista* was in turn 'the model for the colonization of the New World.'

13  Felix Driver, *Geography Militant: Cultures of Exploration and Empire* (Oxford UK: Blackwell, 2001) pp. 200–1, 216.

14  Nicholas Thomas, *Colonialism's Culture: Anthropology, Travel and Government* (Polity Press, 1994) p. 10.

15  Terry Eagleton, *Crazy John and the Bishop and Other Essays* (Notre Dame: University of Notre Dame Press, 1999).

16  Wlad Godzich, *On the Emergence of Prose* (Minneapolis: University of Minnesota Press, 1987) p. 162.

17  David Cannadine, *Ornamentalism: How the British Saw their Empire* (London: Allen Lane, The Penguin Press, 2001).

18  Graham Huggan, *Postcolonial Exotic: Marketing the Margins* (London: Routledge, 2001).

19  Michael Binyon, 'How Islamic world learned to hate the US', *Times* 13 September 2001; President George W. Bush, 'Address to a Joint Session of Congress and the American People', 20 September 2001, at http://www.whitehouse.gov/news/releases/2001/09. See also Ziauddin Sardar, Meryl Wyn Davies, *Why do people hate America?* (Cambridge UK: Icon Books, 2002).

20  Roxanne Euben, 'Killing (for) Politics: *Jihad*, martyrdom and political action', *Political Theory*, 30 (2002) pp. 4–35; cf. Seumas Milne, 'They can't see why they are hated', *Guardian* 13 September 2001.

21  John Wideman, 'Whose war? The color of terrorism', *Harper's Magazine*, March 2002, pp. 33–8; quotations at pp. 36–7.

22  Fareed Zakaria, 'The politics of rage: Why do they hate us?' *Newsweek* 15 October 2001.

23  Cf. Edward Said, *Covering Islam* (New York: Vintage, 1997) p. 8: Within the optic of contemporary Orientalism, '"Islam" seems to engulf all aspects of the diverse Muslim world, reducing all to a special malevolent and unthinking essence.'

24  Mahmood Mamdani, 'Good Muslim, bad Muslim – an African perspective', SSRC/After September 11 at http://www.ssrc.org/sept11/essays; reprinted (with revisions) as 'Good Muslim, bad Muslim: a political perspective on culture and terrorism', in Eric Hershberg and Kevin Moore (eds) *Critical Views of September 11: Analyses from around the world* (New York: The New Press, 2002) pp. 44–60. See also *Islamophobia: A Challenge for Us All* (London UK: Runnymede Trust, 1997) which distinguishes between 'closed' views that treat Islam as (for example) monolithic, static and irredeemably other, and 'open' views that treat Islam as diverse, dynamic and intrinsically interdependent. These criteria do not exempt Islam from disagreement or criticism, but they do identify the well-springs of what the authors see as a 'phobic dread' of Islam.

25  Fareed Zakaria, 'Our way', *New Yorker* 14/21 October 2002, p. 81.

26  Veena Das, 'Violence and translation', SSRC/After September 11 at http://www.ssrc.org/sept11/essays; also in *Anthropological Quarterly* 75 (2002) pp. 105–112; quotation at p. 108.

27  Perry Anderson, 'Internationalism: a breviary', *New Left Review* 14 (March–April 2002) pp. 5–25: 'The American ideology of a republic simultaneously exceptional and universal:

unique in the good fortune of its institutions and endowments, and exemplary in the power of its radiation and attraction' has reached a climax today when 'in the absence of any alternative or countervailing power, American hegemony has for the first time been able to impose its self-description as the global norm'.

28  Arundhati Roy, 'The algebra of infinite justice', *Guardian* 29 September 2001; reprinted in her *Power Politics* (Cambridge MA: South End Press, 2001) pp. 105–124. I should also add that to ask these questions is not to be 'anti-American', an outrageous charge that not only signals a failure of critical intelligence but also betrays the victims of the terrorist attacks. See Simon Schama, 'The dead and the guilty', *Guardian* 11 September 2002; Arundhati Roy, 'Not again', *Guardian* 27 September 2002.

29  Judith Butler, 'Explanation and exoneration, or What we can hear', *Theory and Event* 5 (2002) (4) pp. 1–21; Joan Didion, 'Fixed opinions, or the hinge of history', *New York Review of Books* 16 January 2003.

30  Cf. Chris Toensing, ''The harm done to innocents', *Boston Globe* 16 September 2001; republished as 'Muslims ask: Why do they hate us?' at http://www.alternet.org, 25 September 2001.

31  Edward Said, 'The clash of definitions', in his *Reflections on exile and other essays* (Cambridge MA: Harvard University Press, 2000) pp. 569–90; quotation at p. 577.

32  Said, Orientalism *op. cit.*, pp. 54–9. Orientalism is not a synonym for colonial discourse; there are, in any case, many Orientalisms, but colonial discourse also relied – and continues to rely – on (for example) discourses of tropicality and primitivism to construct its others as 'other'.

33  See Gillian Rose, 'Performing space', in Doreen Massey, John Allen and Phillip Sarre (eds) *Human Geography Today* (Cambridge UK: Polity Press, 1999) pp. 247–259. For this reason Alexander Moore's critique of Said seems to me profoundly mistaken. For Said, space is *not* 'materialized as background' – 'radically concretized as earth' – and it is wrong to use Soja's 'illusion of opaqueness' – space as 'a superficial materiality, concretized forms susceptible to little else but measurement and phenomenal description' – as a stick with which to beat him. For Said too, 'space' is an effect of the practices of representation. Cf. Alexander Moore, 'Postcolonial "textual space": towards an approach', *SOAS Literary Review* 3 (2001) pp. 1–23; Edward Soja, *Postmodern Geographies: The Reassertion of Space in Critical Social Theory* (London: Verso, 1989) p. 7. Soja was trading on Henri Lefebvre, *The Production of Space* (Oxford UK and Cambridge MA: Blackwell, 1991).

34  Cf. Homi Bhabha, *The Location of Culture* (London: Routledge, 1994) p. 219. Judith Butler describes the conditional, creative possibilities of performance as 'a relation of being implicated in that which one opposes, [yet] turning power against itself to produce alternative political modalities, to establish a kind of political contestation that is not a "pure opposition" but a difficult labour of forging a future from resources inevitably impure': *Bodies That Matter* (London and New York: Routledge, 1993) p. 241.

35  Zygmunt Bauman, *Globalization: The Human Consequences* (Cambridge UK: Polity Press, 1998) pp. 75–6, 88–92; idem, *Liquid Modernity* (Cambridge UK: Polity Press, 2000) pp. 117–9.

36  Idem, 'Reconnaissance wars of the planetary frontierland', *Theory, Culture and Society* 19 (2002) pp. 81–90; quotation at p. 82.

37  Don DeLillo, 'In the ruins of the future: reflections on terror and loss in the shadow of September', *Harper's Magazine*, December 2001.

38  Michael Ignatieff, 'The burden', *New York Times* 5 January 2003.

39  John Urry, 'The global complexities of September 11', *Theory, Culture and Society* 19 (2002) pp. 57–69; quotations at pp. 63–4. Urry's argument mobilizes a disconcertingly utopian image of New York City – which certainly has its 'wild zones' – and in mapping

'asylum-seeking' onto the same plane as transparently illegal activities he makes an objectionable contraction of his own.

40 David Harvey, *The Condition of Postmodernity* (Oxford UK and Cambridge MA: Blackwell, 1989); cf. Cindi Katz, 'On the grounds of globalization: a topography for feminist political engagement', *Signs* 26 (2001) pp. 1213–1234, especially pp. 1224–5.

41 See *The Colonial Present*, Chapter 9.

42 Michael Hardt and Antonio Negri, *Empire* (Cambridge MA: Harvard University Press, 2000) pp. 332–3. I have the deepest reservations about this book from which, as Timothy Brennan notes, the 'colonized of today' are evacuated. 'Why', he asks, 'at a moment of American imperial adventure, almost Roman in its excess, is the end of imperialism confidently announced?'. He continues: 'As the United States ushers a new government into Kabul, criminalizes those who quibble with its military plans, and thumbs its nose at the Geneva conventions, *Empire*'s thesis that imperialism has ended is likely to seem absurd.' And yet Brennan's critique is all the more cogent because it focuses – unwaveringly – on the epistemological flaws in the edifice: 'The Empire's new clothes', *Critical Inquiry* 29 (2003) pp. 337–67.

43 Hardt and Negri, Empire *op. cit.*, pp. 188, 194–5; Bauman, 'Reconnaissance' *op. cit.*, pp. 83–4.

44 Cf. Katharyne Mitchell, 'Different diasporas and the hype of hybridity', *Environment and Planning D: Society & Space* 15 (1997) pp. 533–53.

45 Das, 'Violence and translation' *op. cit.*, p. 112 n.3.

46 See Gregory, *The Colonial Present op. cit.*, Chapter 9.

47 See Gregory, *The Colonial Present op. cit.*, Chapter 9.

48 See Derek Gregory, 'Scripting Egypt: Orientalism and cultures of travel', in James Duncan and Derek Gregory (eds) *Writes of passage: reading travel writing* (London: Routledge, 1999) pp. 114–150.

49 John Urry, *The Tourist Gaze*, (London: Sage 2002) p. 74.

50 See http://www.whitehouse.gov/news/releases/2001/09/20010927-1.html.

51 Gary Younge, 'Under a veil of deceit', *Guardian* 19 August 2002.

52 Homi Bhabha, 'Double visions', *Artforum* 30.5 (1992) pp. 82–90; quotation at p. 88.

53 Suvendri Perera, 'What is a camp', *Borderlands* 1 (2002) at http://www.borderlandsejournal.adelaide.edu.au.

54 Giorgio Agambem, *Homo Sacer: Sovereign Power and Bare Life* (Stanford CA: Stanford University Press, 1998; first published in Italian in 1995): 'The exception is that which cannot be included in the whole of which it is a member and cannot be a member of the whole from which it is always already included.' Cf. Edward Soja, *Thirdspace: Journeys to Los Angeles and Other Real-and-Imagined Places* (Oxford: Blackwell, 1996).

55 Georgio Agamben, 'Über Sicherheit und Terror', *Frankfurter Allgemeine Zeitung* 20 September 2001, translated as 'Security and terror', *Theory and Event* 5:4 (2002); see also Michael Shapiro, 'Wanted, dead or alive', *loc. cit.*, and Bülent Diken, Carsten Basse Laustsen, 'Zones of indistinction: security, terror and bare life', *Space and Culture* 5 (2002) pp. 290–307.

56 Human Rights Watch, *Presumption of Guilt: Human Rights Abuses of post-September 11 Detainees*, at http://www.hrw.org, August 2002; 'For whom the Liberty Bell tolls', *Economist* 29 August 2002; see also David Cole, 'Enemy aliens and American freedoms', *The Nation*, 23 September 2002.

57 Paul Passavant and Jodi Dean, 'Representation and the event', *Theory and Event* 5/4 (2002).

58 Kanishka Jasuriya, '9/11 and the new "anti-politics' of "security"'', SSRC/After September 11 at http://www.ssrc.org/sept11/essays; reprinted as 'September 11, security and the new post-liberal politics of fear', in Hershberg and Moore (eds) Critical views *op. cit.*,

pp. 131–147; see also his 'The exception becomes the norm: law and regimes of exception in East Asia', *Asia-Pacific Law and Policy Journal*, 2 (2001). Human Rights Watch has drawn special attention to the use of the 'war on terrorism' as a cloak for repression in Australia, Belarus, China, Egypt, India, Israel, Jordan, Kyrgyzstan, Macedonia, Malaysia, Russia, Syria, the USA, Uzbekistan and Zimbabwe: Human Rights Watch, *Opportunism in the face of tragedy: repression in the name of terrorism*, at http://www.hrw.org, September 2002.

59  Oren Yiftachel, 'The shrinking space of citizenship: ethnocratic politics in Israel', *Middle East Report* 223 (Summer 2002); Neve Gordon, 'The enemy within', in Carey and Shainin (eds) Other Israel *op. cit.*, pp. 99–105; quotation at p. 102; Jonathan Cook, 'Unwelcome citizens of a racist state', *Al-Ahram*, 26 September–2 October 2002.

60  Michael Dibdin, *Dead Lagoon* (London: Faber, 1994) pp. 113–15, 312.

61  Michael Ignatieff, *The Needs of Strangers* (London: Chatto and Windus, 1984); see also Stuart Corbridge, 'Marxism, Modernities and Moralities: Development Praxis and the Claims of Distant Strangers', *Environment and Planning D: Society and Space* 11 (1993) 449–72; idem, 'Development ethics: distance, difference, plausibility', *Ethics, Place and Environment* 1 (1998) 35–54.

62  Cf. Ash Amin, 'Spatialities of globalization', *Environment and Planning* 34 (2002) 385–399; Eric Sheppard, 'The spaces and times of globalization: place, scale, networks and positionality', *Economic Geography* 78 (2002) pp. 307–30.

# Index